Advances in Anatomy
Embryology and Cell Biology

Vol. 74

Editors
F. Beck, Leicester W. Hild, Galveston
J. van Limborgh, Amsterdam R. Ortmann, Köln
J.E. Pauly, Little Rock T.H. Schiebler, Würzburg

Joseph Altman   Shirley A. Bayer

Development of the Cranial Nerve Ganglia and Related Nuclei in the Rat

With 64 Figures

Springer-Verlag
Berlin Heidelberg New York 1982

Joseph Altman
Laboratory of Developmental Neurobiology
Department of Biological Sciences
Purdue University
West Lafayette, Indiana 47907
U.S.A.

Shirley Bayer
Laboratory of Developmental Neurobiology
Department of Biological Sciences
Purdue University
West Lafayette, Indiana 47907
U.S.A.

ISBN-13: 978-3-540-11337-9   e-ISBN-13: 978-3-642-68479-1
DOI: 10.1007/978-3-642-68479-1

Library of Congress Cataloging in Publication Data
Altman, Joseph, 1925–   Development of the cranial nerve ganglia and
related nuclei in the rat. (Advances in anatomy, embryology, and cell
biology; v. 74) Includes bibliographical references and index. 1. Nerves,
Cranial. 2. Rats - - Developmental. 3. Developmental neurology.
4. Embryology - - Mammals. I. Bayer, Shirley A. (Shirley Ann), 1940–   .
II. Title. III. Series. [DNLM: 1. Cranial nerves - - Embryology. 2. Cranial
nerves - - Growth and development. 3. Brain - - Cytology.
W1 AD433K v. 74 / WL 330 A468d] QL801.E67 vol. 74 [QL931]
574.4s 81-23259
ISBN 0-387-11337-1 (U.S.) [559.03'3] AACR2

Composition: Schreibsatz Service Weihrauch, Würzburg
Printing and binding: H. Stürtz AG, Universitätsdruckerei, Würzburg
2121/3321-543210

Contents

Acknowledgments

This work was supported by grants from the National Institutes of Health and the National Science Foundation. We wish to thank T. Chen, S. Evander, C. Landon, and K. Shuster for their excellent technical assistance.

1 Introduction

The aim of this investigation is threefold: (a) to determine the time of origin of neurons of the rat cranial nerve ganglia; (b) to reexamine the embryonic development of the cranial nerve ganglia in the light of these dating results; and (c) to attempt to relate the chronology of these peripheral events to developmental events in those nuclei of the medulla that are intimately associated with the cranial nerve ganglia.

Although thymidine-radiography has been used for over 2 decades to investigate the time of origin of neurons, most of these studies dealt with central nervous structures. There are relatively few studies available concerning the birth dates of neurons in the peripheral nervous system. In fact, to our knowledge, there is only a single thymidine-radiographic report available dealing with the time of origin of neurons of a cranial nerve ganglion in a mammal; this is the recent study by Forbes and Welt (1981) of neurogenesis in the trigeminal ganglion of the rat. In the present study we determined the birth dates of neurons of the trigeminal, facial, vestibular, glossopharyngeal, and vagal ganglia of the rat. We utilized the progressively delayed comprehensive labeling procedure, a method which, in contrast to the single-pulse labeling procedure, allows the exact quantification of the proportion of neurons formed on a particular day. While we have made reference to the time of origin of spiral ganglion cells in relation to the vestibular ganglion cells, a quantitative study of temporal differences in neuron production along the cochlea was not attempted here.

The study of the embryonic development of the cranial nerve ganglia began a little over a century ago, and the old literature was critically reviewed by Adelmann (1925). One of the classical concerns was the question whether the cranial nerve ganglia are derived from the neural crest, the placodes, or both. Another concern was the number and identity of the placodes from which the ganglia are derived; for instance, are the facial and acoustic ganglia derived from a single primordium (Weigner 1901) or from separate primordia (Streeter 1906)? Our reexamination of the embryonic development of the cranial nerve ganglia was motivated by the expectation that information about the time of origin of neurons of the different ganglia, and differences between and within these ganglia, would aid us in identifying their germinal sources and fates. To facilitate this attempt further, we utilized, in addition to conventional paraffin-embedded embryos, a large series of embryos embedded in methacrylate. This material has helped us to distinguish young neurons from primitive cells and to visualize in the same material neurites and axons.

Our third aim was to relate the development of the cranial nerve ganglia and nerves to the development of the sensory and motor nuclei of the medulla that are directly associated with these cranial nerves. This attempt deals with the theoretically important issue of the temporal relations between central and peripheral events in the establishment of the circuitry of the nervous system. We utilized for this aspect of the study the embryonic materials referred to and leaned heavily on our recent datings of the time of origin of neurons in the major nuclei of the lower and upper medulla (Altman and Bayer 1980a–d). Because the study focuses on the cranial nerve ganglia, we did not deal here with the development of pure motor nuclei (the oculomotor, trochlear, abducent, and spinal accessory nuclei). For reasons that will be described later, we have briefly considered the hypoglossal nucleus.

1

2 Materials and Methods

2.1 Whole-Body Thymidine Radiograms

Purdue-Wistar pregnant rats were injected subcutaneously with two successive daily doses of ^{3}H-thymidine at the following gestational ages: E11+12, E12+13, E13+14, E14+15, E15+16, E16+17, and E17+18. The day of sperm positivity was counted as day E1; the specific activity of the radiochemical was 6.0 Ci/mM; injections of 5 μCi/g body w. were made between 9:00 and 11:00 a.m. Two dams from each injection group were anesthetized on day E21 and their fetuses were removed. The bodies of all of the fetuses were blocked coronally, and their heads in one of the three standard planes, and immediately immersed in Bouin's solution. After 24 h of fixation, the blocks were embedded in either paraffin or methacrylate. The paraffin specimens were sectioned serially at 6 μm, and the methacrylate specimens at 3 μm, and every tenth section was saved. Successive sections were either stained with cresyl violet and hematoxylin-eosin for examination without nuclear emulsion, or prepared for autoradiography. The latter were coated with Kodak NTB-3 emulsion in the dark, exposed for 90 days with a dessicant, developed in Kodak D-19, and stained with hematoxylin-eosin.

Autoradiograms from the paraffin-embedded collection (Fig. 1) were used for quantitative purposes, following the same procedure as was used in previous studies of this series (e.g., Altman and Bayer 1980a). The proportion of labeled to unlabeled neurons was determined in the cranial nerve ganglia at × 625 magnification. In all instances a minimum of 100 (up to several hundred) cells were classified in each ganglion per animal. The estimation of the proportion of cells produced (ceasing to divide) on a particular day was based on the progressively delayed comprehensive labeling procedure. The rationale of this procedure is that as long as virtually all the cells of a selected ganglion can be labeled (in the populations studied here 100% of labeling can be accomplished with two successive daily injections on days E11 and E12), all the cells are considered to be proliferating precursors that have not started to differentiate. When with delayed onset of injections the total population can no longer be tagged, the proportion of cells that is not tagged as a result of delay by a single day is taken to be the complement that differentiated on the previous day. As an example, the proportion of cells generated on day E13 is determined as follows: E13 = (labeled cells on days E13+14) − (labeled cells on days E14+15). Autoradiograms from the methacrylate-embedded collection were used for qualitative assessments.

2.2 Paraffin- and Methacrylate-Embedded Embryos

The paraffin-embedded material was the same as that which we analyzed in previous studies (e.g., Altman and Bayer 1980a). This collection consists of a large number (over 200) of embryos and fetuses, ranging in age at daily intervals from day E10 to E22. The material was cut in the three standard planes at 6 μm, and all sections were saved in embryos aged E10−E14, every fifth section at ages E15−E16, and every tenth section at ages E17−E22. Alternate sections were stained with cresyl violet and hematoxylin-eosin. The methacrylate-embedded collection is similar to the above except for the processing of the tissue, and that the sections were cut at 3 μm. Over 140 speci-

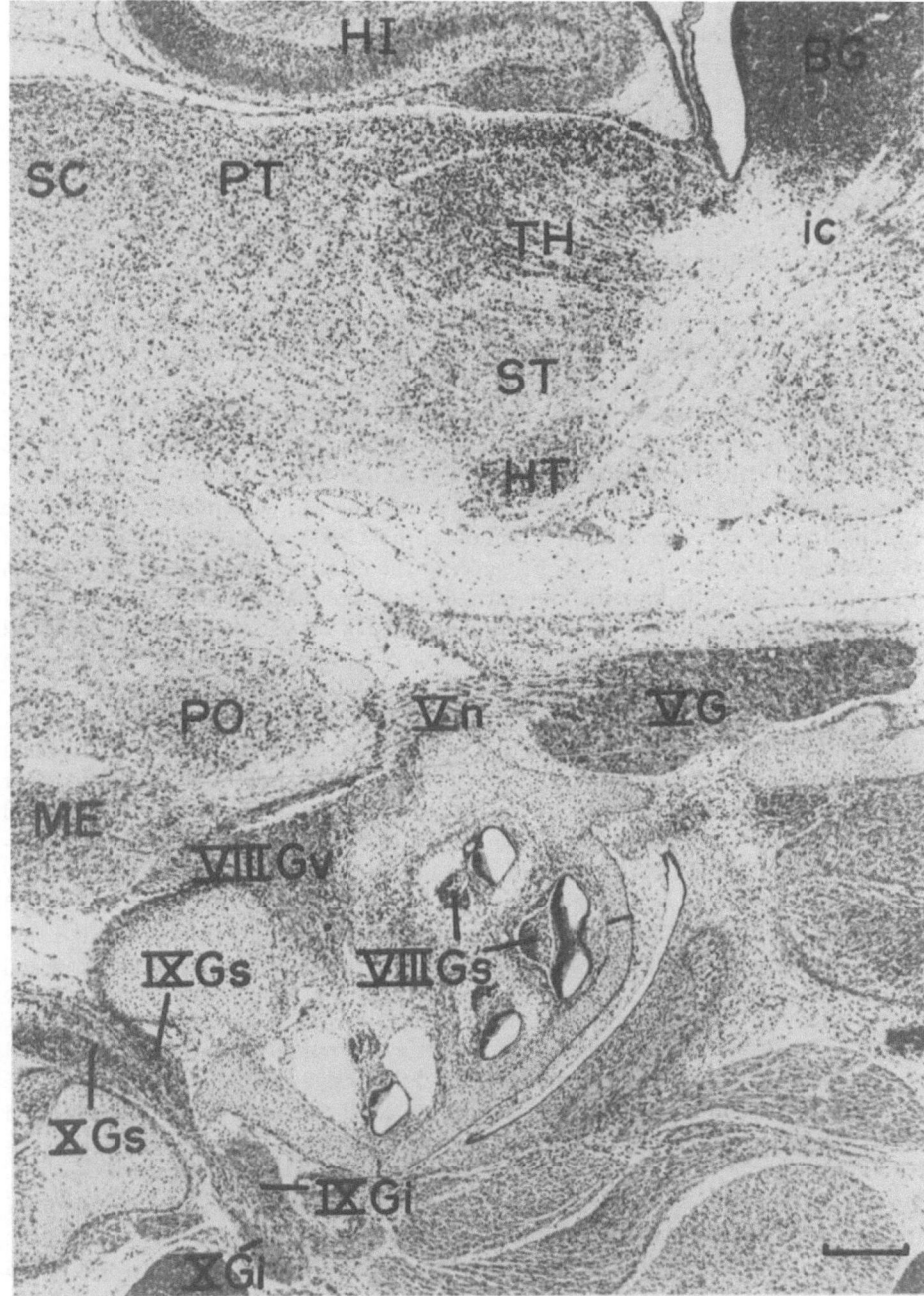

Fig. 1. Parasagittal low-power thymidine radiogram from a rat labeled on days E14+15. Portions of all cranial nerve ganglia, except the facial, are illustrated. Paraffin embedding. *Scale:* 300 μm

mens were available between the ages of E13 and E22. This newer material not only provides better cytological preservation, but also allows visualization of fiber tracts and, under optimal conditions, single neurites or axons.

3 Development of the Trigeminal Ganglion in Relation to the Trigeminal Nuclei

3.1 Time of Origin of Trigeminal Ganglion Cells

Background. The trigeminal (semilunar, gasserian) ganglion contains the perikarya of primary sensory neurons that are the source of the peripherally and the centrally directed fibers of the trigeminal nerve. The peripheral (distal) fibers form three branches, the ophthalmic, the maxillary, and the mandibular nerves, and these provide afferents to components of the face and the head that are considered to be derivatives of the first branchial arch. The central (proximal) fibers enter the rostral medulla in a ventrolateral position and bifurcate into locally terminating and descending branches. The trigeminal ganglion is a prominent structure in most vertebrates, and it is usually said to be composed of two lobes. For instance, in the chick, D'Amico-Martel and Noden (1980) distinguished an anterior lobe related to the ophthalmic nerve, and a posterior lobe related to the maxillary and mandibular nerves. In the cat, Allen (1924) described a large anterior and medial portion, consisting of ophthalmic-maxillary cells, and a smaller posterior and lateral division made up of mandibular cells. More precise delineation of the projections of trigeminal ganglion cells has been attempted by several investigators with classical and modern tracing techniques (Kerr and Lysak 1964; Mazza and Dixon 1972; Gregg and Dixon 1973; Arvidsson 1975; Noden 1980). As in other sensory ganglia, large and small neurons have been distinguished in the trigeminal ganglion; these two types have also been characterized as light and dark cells, respectively (see below). The time of origin of neurons of the trigeminal ganglion has recently been studied in the chick (D'Amico-Martel and Noden 1980) and the rat (Forbes and Welt 1981). Forbes and Welt used the single-pulse labeling procedure and found that heavily labeled large neurons were present in highest concentration on day E12 and small neurons on day E13.

Results. The trigeminal ganglion of the rat has an asymmetrical dumbbell shape and can be subdivided into a larger anteromedial lobe and a smaller posterolateral lobe (Fig. 2). In the animals labeled on days E11+12, all the cells of anteromedial and posterolateral lobes were lightly labeled. Heavily labeled large cells began to appear in the E12+13 group and a few of the large neurons were unlabeled. There was no preferential location for either the heavily labeled or unlabeled cells. A fair proportion of the neurons were heavily labeled in the E13+14 group. In the E14+15 group the highest proportion of remaining labeled neurons were of the small type (Fig. 3), and in the E15+16 animals the few remaining labeled neurons were exclusively small. In the animals injected on E16+17 or E17+18 (Fig. 4), none of the neurons but all of the satellite cells were labeled. Our cell counts (Fig. 5) showed that the neurons of both lobes of the trigeminal ganglion are generated between days E11 and E15, with 73% of the anteromedial neurons and 65% of the posteromedial neurons being generated on days E13 and E14.

4

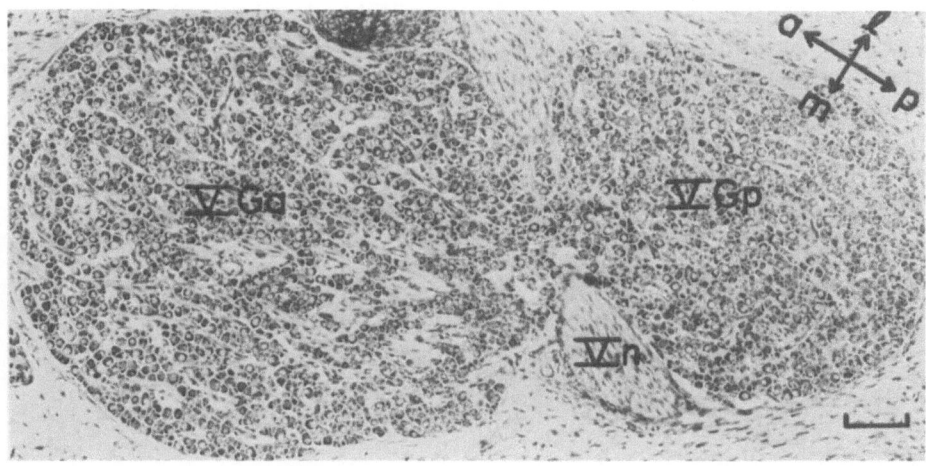

Fig. 2. Horizontal section through the trigeminal ganglion of a day E19 rat. Orientation with respect to the midline indicated in the *upper right hand corner*. Methacrylate embedding; cresyl violet. *Scale:* 100 μm

Comments. Our results, indicating a protracted time span of neuron production in the trigeminal ganglion from day E11 to day E15, are in agreement with Forbes and Welt's (1981: Fig. 5) findings based on the single-pulse labeling technique. However, our quantitative findings do not agree with their conclusion that peak production of trigeminal neurons occurs on day E12, which implies that relatively few trigeminal ganglion cells should be labeled with injections of ^{3}H-thymidine started on day E13. Rather, we found that 83% of the neurons could still be labeled in the E13+14 group in both lobes of the trigeminal ganglion. In general, there is little resemblance between the daily distribution of heavily labeled cells reported by Forbes and Welt (1981: Fig. 5) and the distribution that we have obtained of the proportion of cells that could no longer be labeled with the progressively delayed comprehensive procedure. We believe that the counting of heavily labeled cells is not an adequate criterion for the quantitative assessment of the proportion of cells generated on a particular day, especially when heavy labeling is not actually obtained and the criterion of what is a "heavily" labeled cell is changed from animal to animal, as was the case in the study of Forbes and Welt.

Our observations indicated that the neurons generated on day E14 were predominantly of the smaller type and those produced on day E15 were exclusively small neurons. This sequence from large to small ganglion cells is in agreement with the observations of Forbes and Welt (1981) in the rat trigeminal ganglion, and with reports by others in the dorsal root ganglia of rats (Lawson et al. 1974; Theisen 1979) and mice (Sims and Vaughan 1979). Our observation of a similar temporal pattern of neuron production in the two lobes of the trigeminal ganglion suggests that they are derived from a shared germinal source, contrary to older ideas (reviewed by Hamburger 1961) that one lobe may be of placodal origin and the other of neural crest derivation. In a recent thymidine-radiographic study (D'Amico-Martel and Noden 1980) a

5

distal-to-proximal cytogenetic gradient was described in the trigeminal ganglion of the chick. In agreement with the observations of Forbes and Welt (1981), we could not detect a cytogenetic gradient in the trigeminal ganglion of the rat.

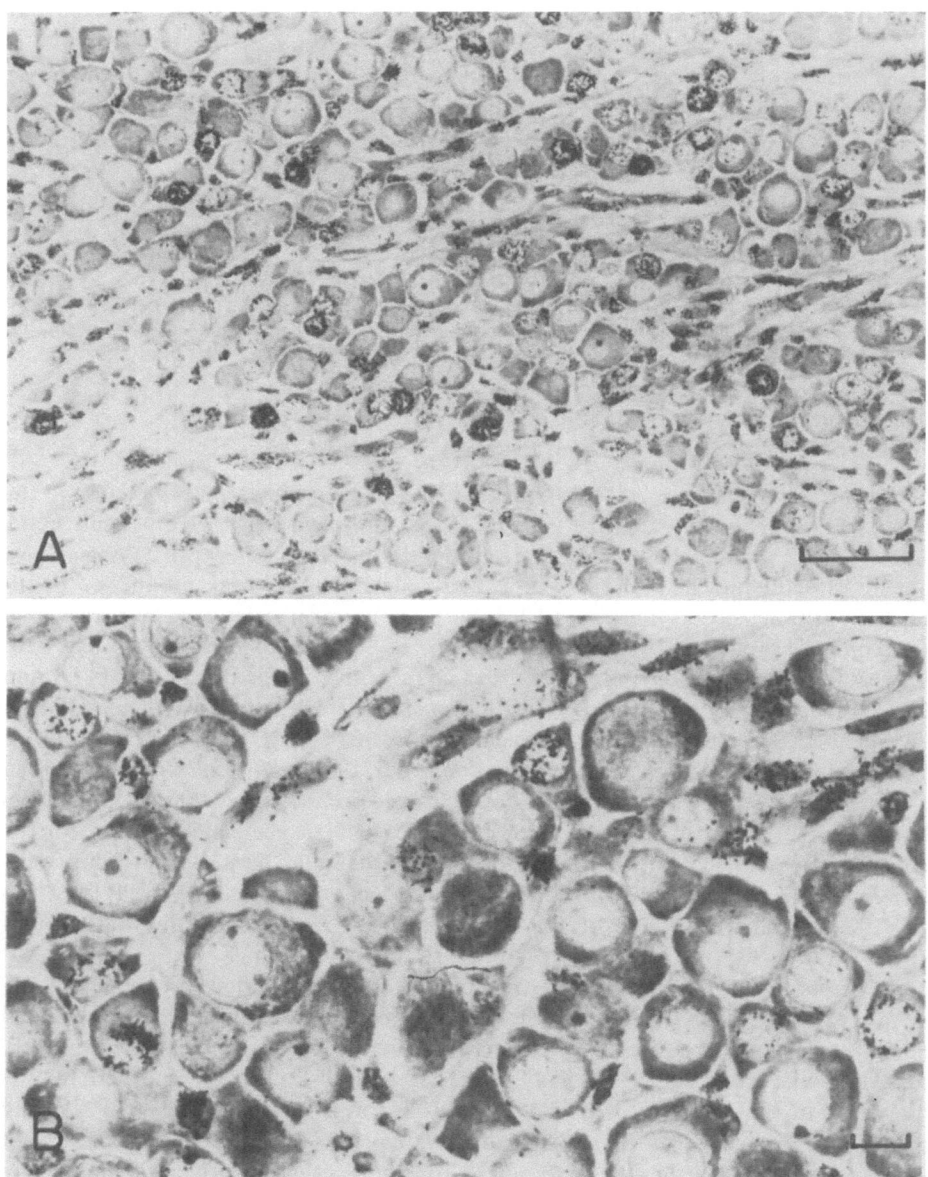

Fig. 3. *A* Labeled small neurons in the trigeminal ganglion of a rat labeled with ^{3}H-thymidine on days E14+15. *B* Trigeminal ganglion cells at higher magnification. Methacrylate; poststaining with cresyl violet. *Scales: A*, 50 μm; *B*, 10 μm

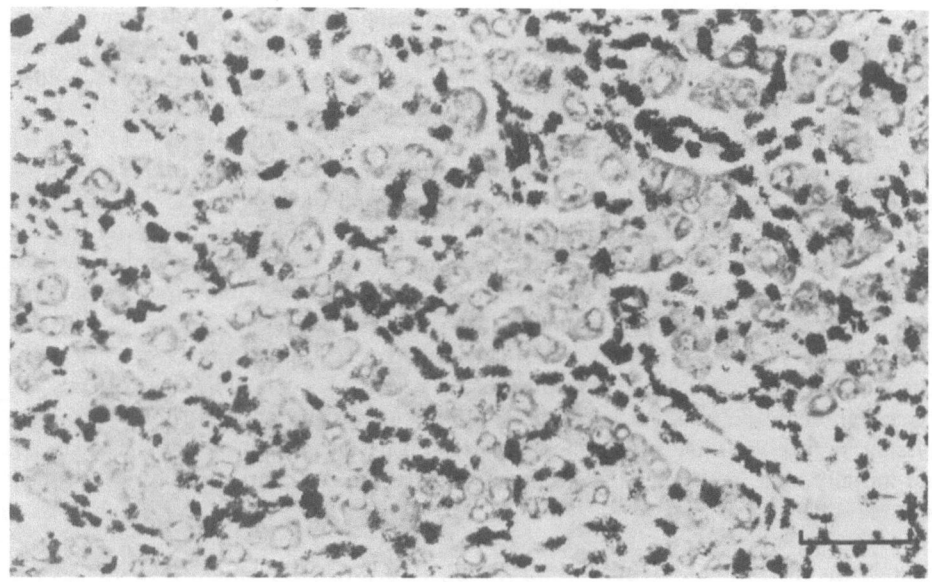

Fig. 4. Thymidine radiogram of the trigeminal nucleus from rat labeled on days E 17+18. Labeling restricted to satellite cells. Paraffin embedding. *Scale:* 50 µm

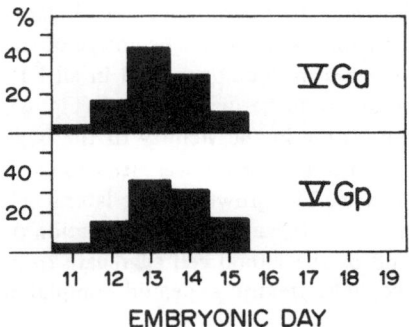

Fig. 5. Time of origin of neurons in the two lobes of the trigeminal nucleus

3.2 Fate of the Two Components of the Trigeminal Anlage

Background. The embryonic development of the trigeminal ganglion has been studied in several species and the old literature has been critically reviewed by Adelmann (1925). Adelmann studied the embryonic development of the cranial nerve ganglia in rats and reported that the proliferating anlage of the trigeminal ganglion is recognizable in five-somite embryos in the vicinity of the neural plate along the territory of the prospective midbrain and "rhombomere A_1." In eight-somite embryos he could trace the ganglion to the mesenchymal condensation of the mandibular arch. This observation suggests that the proliferation of cells destined to form the trigeminal ganglion is in progress by day E11 (if the day of sperm positivity is counted as day E1), which is in agreement with the observations of Angulo (1951). On the basis of his observations,

7

Adelmann (1925) concluded that the entire trigeminal ganglion derives from the neural crest. In contrast, other investigators (for instance, Bartelmez 1922) suggested that the trigeminal ganglion is derived from both the neural crest and the regional epidermal placode. On the basis of experimental work in the chick, Hamburger (1961) proposed that the small and large neurons of the ganglion are selectively derived from the neural crest and the placode. This conclusion has been supported by others (Johnston 1966; Noden 1975), although the regional segregation of small and large neuros disappears as development proceeds (Gaik and Farbman 1973b; D'Amico-Martel and Noden 1980). The validity of the characterization of small trigeminal ganglion cells as dark cells and the larger ones as light cells has been debated for some time. While it has been claimed that in man (Beaver et al. 1965; Moses 1967), cat (Pineda et al. 1967), and monkey (Carmel and Stein 1969) dark cells are fixation artefacts, studies in the chick (Gaik and Farbman 1973a) and the rat (Peach 1972) concluded that the two cell types genuinely differ from each other in terms of density and distribution of cytoplasmic organelles. But, to complicate matters, it has been claimed that the small cells are initially light in the chick and only become dark in the course of maturation (Gaik and Farbman 1973b).

Results. In the rat, neurulation occurs on day E10. Beginning on day E11, the growth of the primitive nervous system proceeds with great rapidity. In the E11 group we were able to distinguish several developmental stages, even though the individual embryos could have differed in actual age by no more than 16 h (the breeding males and femals were kept together from 4:00 p.m. to 8:00 a.m.). The youngest embryo was presomitic, with a minimum of cephalic flexion, and the oldest embryo had reached the 11-somite stage; in it the cephalic flexion was quite advanced. The anlage of the trigeminal ganglion was recognizable in an E11, three-somite embryo, and in all E11 embryos two cell aggregates, a medial and a lateral, could be distinguished (Fig. 6). The medial cell aggregate was apposed to the neural tube in the vicinity of the large rostral rhombomere (rhombomere 1), while the lateral cell aggregate was situated close to the ectodermal thickening rostral to the otic vesicle. The growth of the lateral cell aggregate was pronounced by day E12 (Fig. 7). In horizontal sections the derivation of the medial cell aggregate from the neural crest, and of the lateral cell aggregate from the placode, was quite suggestive (Fig. 7). The cephalic flexure appeared completed by day E12.

In day E13 embryos the cells of the lateral anlage began their morphological differentiation. This was more evident in E14 embryos (Fig. 8A, B), in which many of the young neurons of the lateral anlage of the trigeminal ganglion began to develop a densely staining, eccentrically located cytoplasmic cap. In some E14 embryos (Fig. 8A), the anteromedial and posterolateral lobes of the trigeminal ganglion could already be distinguished. The difference between small and large differentiating neurons began to be recognizable in day E15 embryos (Fig. 9A) and this was quite evident in the E16 group (Fig. 9B). The differentiating small and large neurons were àt no time spatially segregated within the growing trigeminal ganglion and there was no difference in their stainability throughout development (Fig. 9). There was a progressive growth in the size of ganglion cells throughout the fetal period, but neurons did not assume a mature appearance until after birth (Fig. 9F). Satellite cells began to appear in the trigeminal ganglion in day E16 embryos (Fig. 9B). At least some of these must have been derived from local proliferation, since the observed mitotic cells could not be neuron

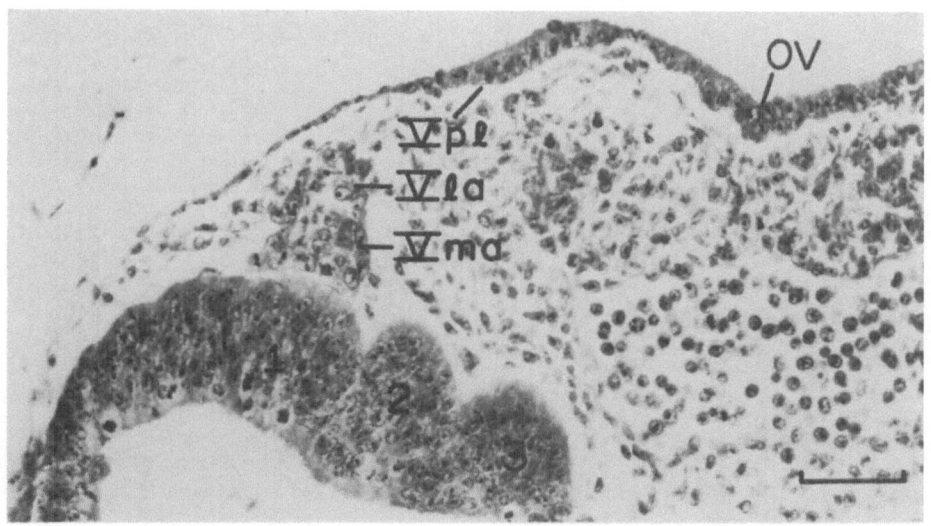

Fig. 6. Horizontal section through a day E11 rat with the two cell aggregates of the trigeminal anlage between rhombomere 1 and the trigeminal placode. *Numbers* refer to the rhombomeres. Paraffin. *Scale:* 50 μm

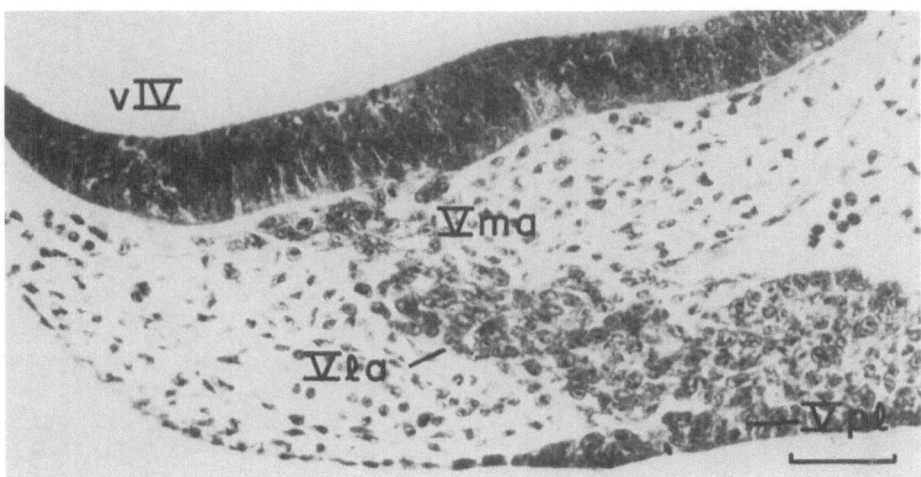

Fig. 7. Horizontal section through the upper medulla of a day E12 embryo to show the relation of the medial and lateral cell aggregates to the neural crest region and trigeminal placode, respectively. Paraffin. *Scale:* 50 μm

precursors, as these ceased to proliferate on day E15 (Fig. 5). The proportion of satellite cells to neurons was still low in the day E19 fetuses (Fig. 9C), but the balance shifted in their favor thereafter (Fig. 9).

In contrast to these progressive morphological changes in the cells of the lateral aggregate of the peripheral trigeminal anlage, which became organized into the two

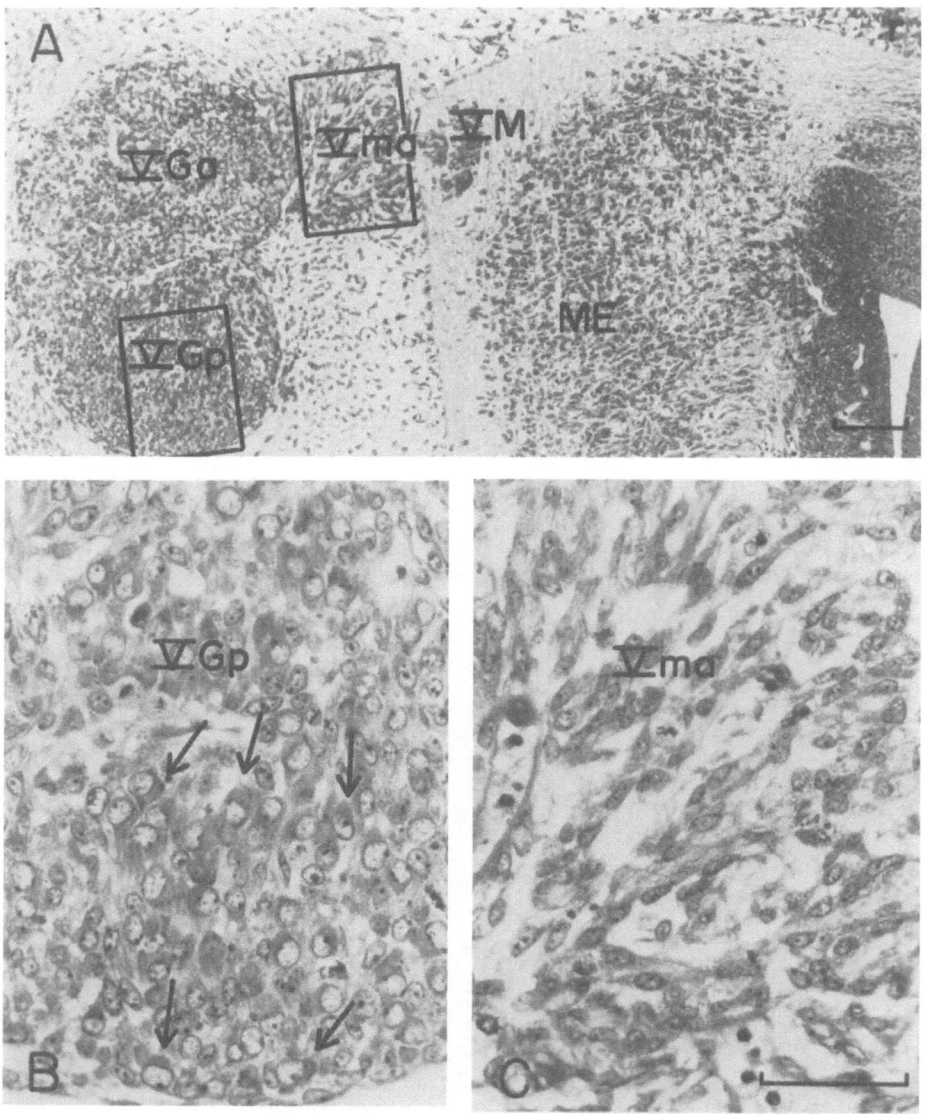

Fig. 8. *A* Horizontal section through the two lobes of the trigeminal ganglion (which apparently derive from the lateral trigeminal anlage) and the separate medial anlage. *B* Many cells in the trigeminal ganglion have eccentric cytoplasmic caps *(arrows)*. *C* There is no sign of morphological cell differentiation in the medial anlage. Methacrylate. *Scales: A* 100 μm; *B* 50 μm

lobes of the trigeminal ganglion (Fig. 8A), the cells derived from the medial aggregate (Fig. 8C) remained small throughout the period of prenatal development. This cell group was traversed by the first contingent of trigeminal afferents on day E13 (Fig. 10A). By day E15 (Fig. 10B), the cells formed a cap-like structure at the boundary of the medulla, where trigeminal afferents penetrated the medulla and formed a band of bi-

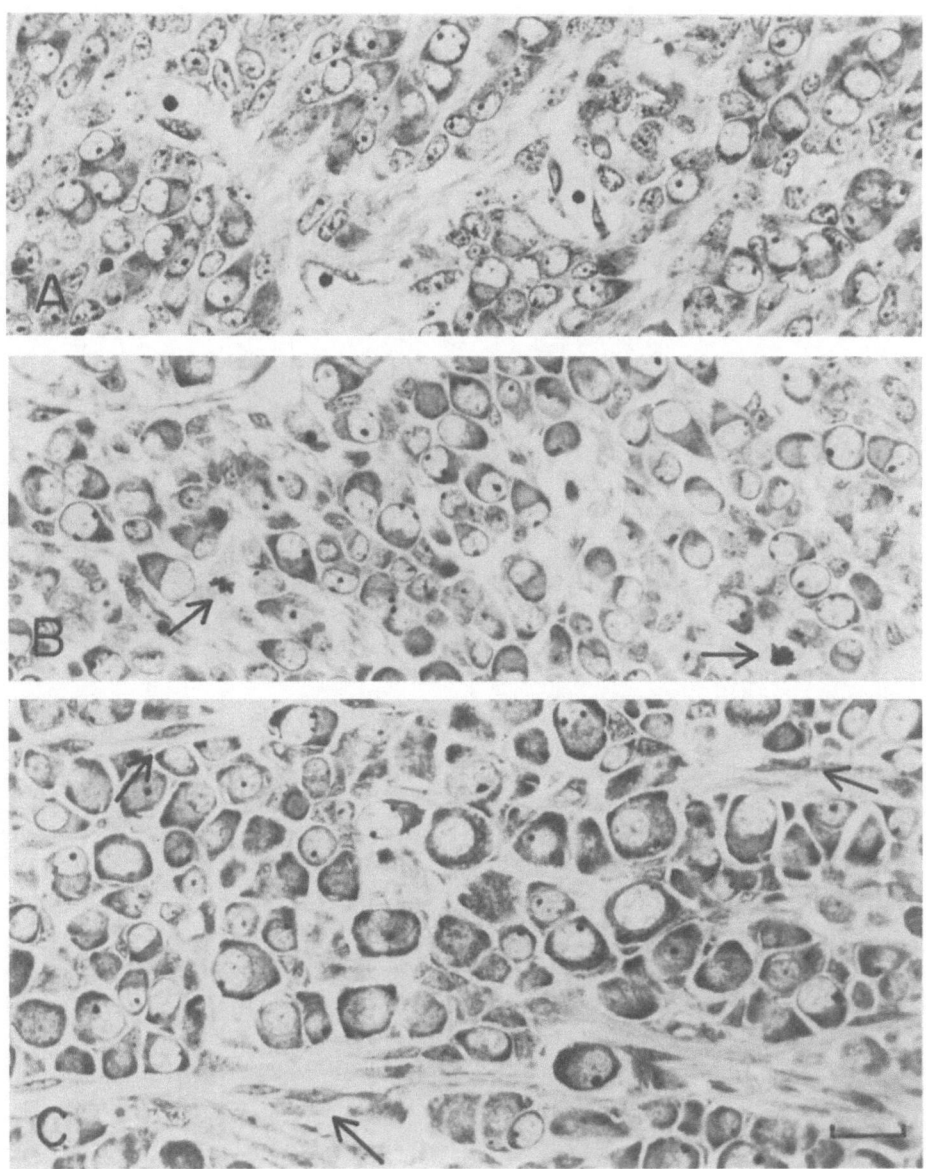

Fig. 9 A–F. Growth and differentiation of neurons of the trigeminal ganglion in embryos of the following ages: E15 *(A)*; E16 *(B)*; E19 *(C)*; E20 *(D)*; E22 *(E)*, and P5 *(F)*. Differences between small and large neurons may appear as early as day E15. The mitotic cells in day E16 embryos *(arrows in B)*, which is after cessation of the generation of ganglion cells, must signal the onset of production of satellite cells. Satellite cells increase in number from day E19 *(arrows in C)* onward. As late as P5 *(F)* there is no detectable difference in the stainability of large and small neurons. Methacrylate. *Scale:* 20 μm

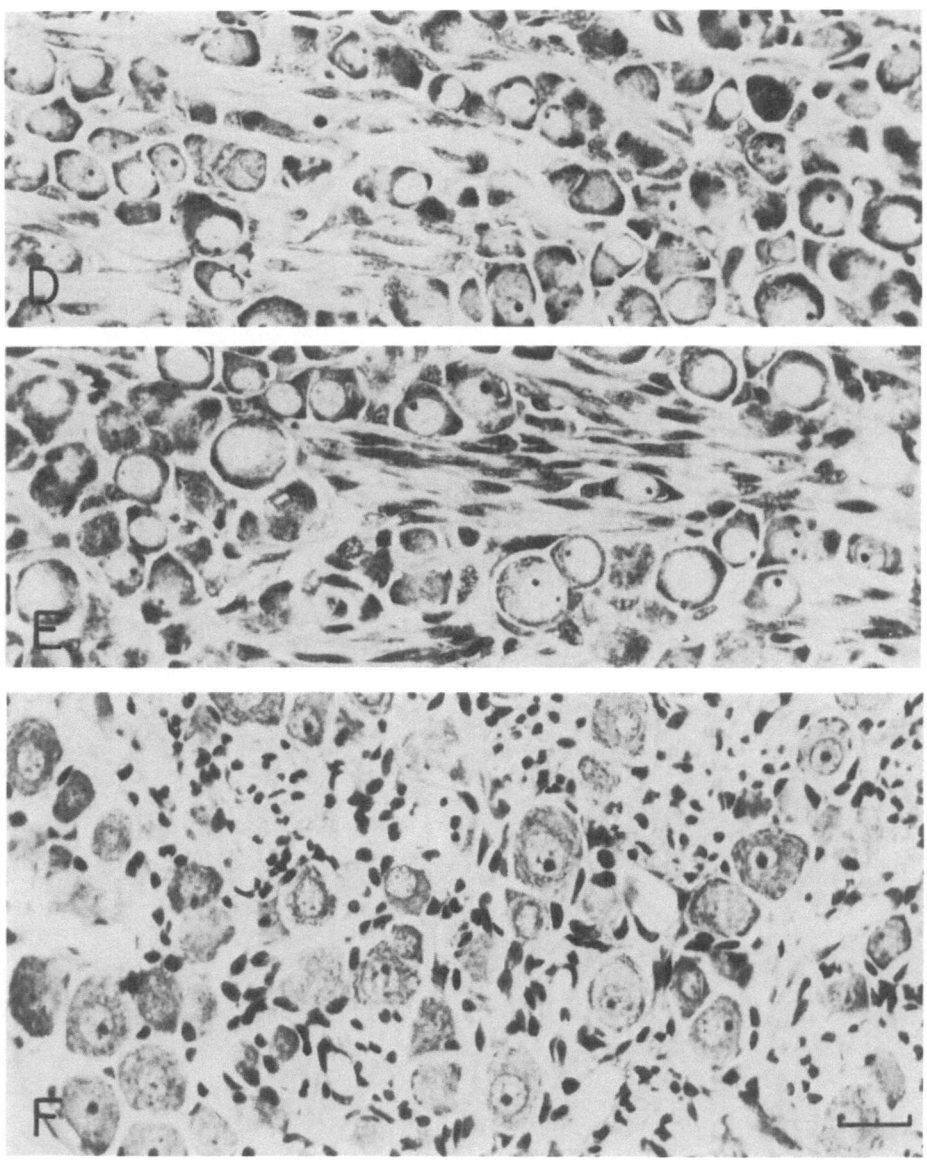

Fig. 9 D–F.

furcating ascending and descending fibers. We shall refer to this transient derivative of the medial cell aggregate as the trigeminal boundary cap. In day E16 embryos (Fig. 11), the primitive spindle-shaped cells seemed to move in large numbers from the boundary cap peripherally along fascicles of the trigeminal nerve, but few, if any, cells had entered the medulla along the central branch of the trigeminal nerve. This process continues throughout the prenatal period, while the trigeminal boundary cap is gradually diminishing in size. The boundary cap was still present on the day before birth,

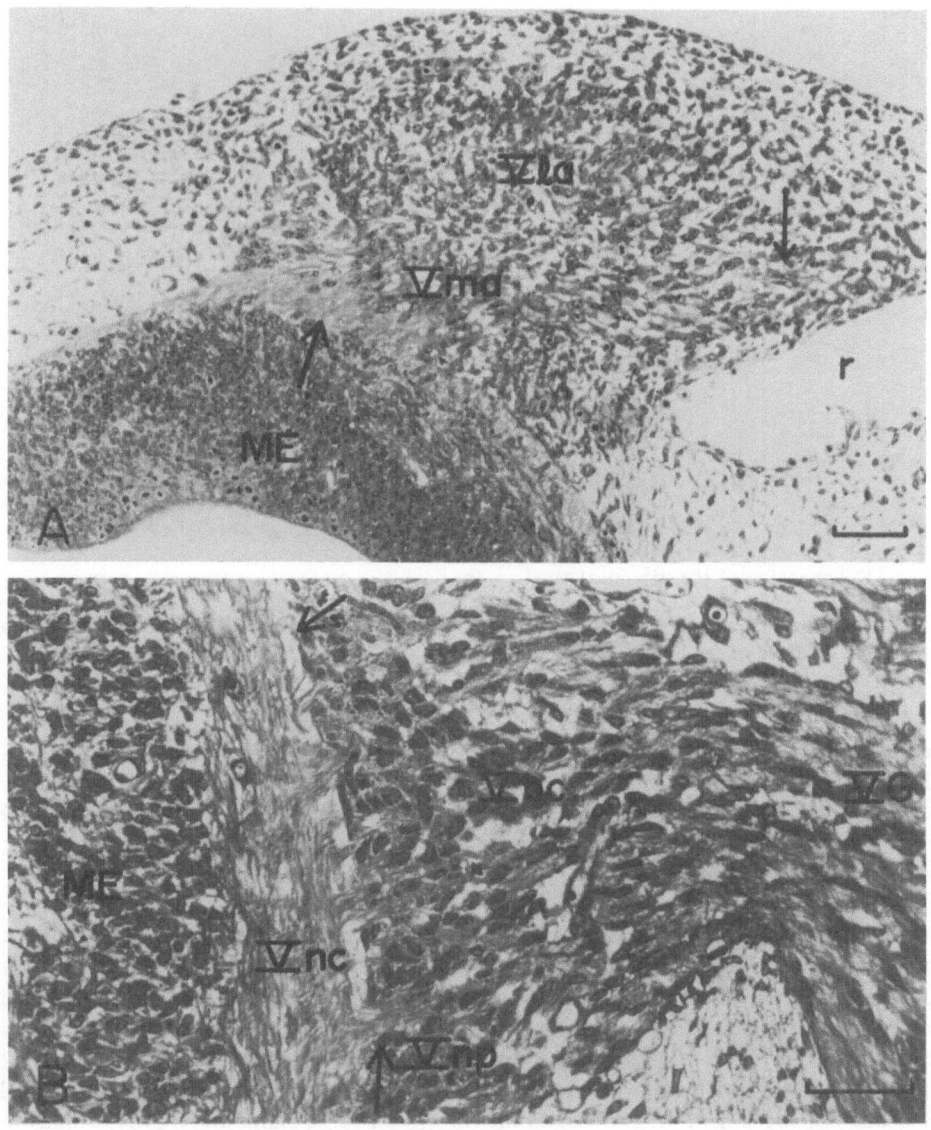

Fig. 10. *A* Horizontal section through the upper medulla of a day E13 embryo. Proximal trigeminal afferents penetrate the medulla *(arrow on left)*, and distal afferents *(arrow on right)* appear to leave the ganglion. *B* In this day E15 embryo the bifurcation of the ascending and descending branches of central trigeminal axons is suggestive. *Arrows* indicate boundary of medulla. Methacrylate. *Scales: A*, 100 μm; *B*, 50 μm

but it was no longer identifiable in P5 rats (the only postnatal, whole-head preparation presently available to us). From these observations it is concluded that the lateral cell aggregate of the trigeminal anlage (which is presumably of placodal origin) is the sole source of neurons of the trigeminal ganglion. The medial trigeminal anlage, which we have distinguished as the transient boundary cap (and which may be of neural crest

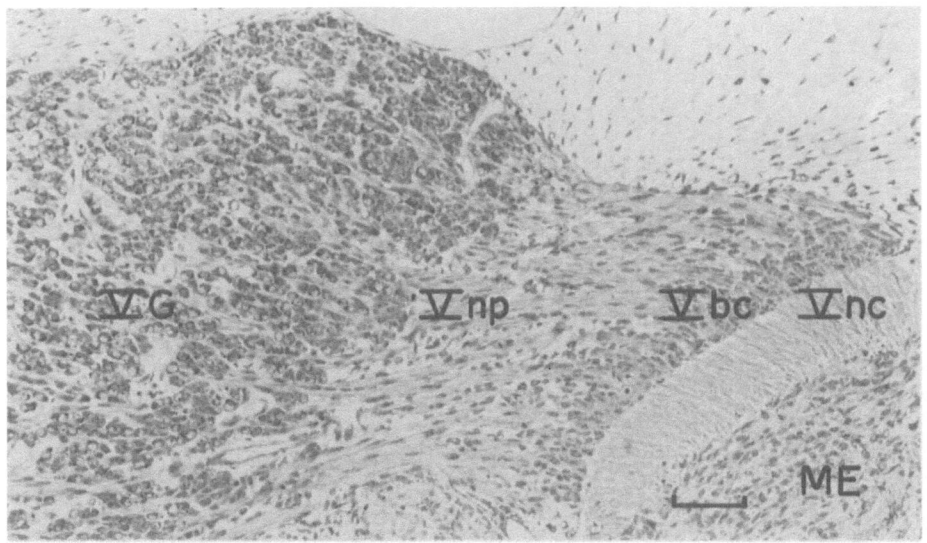

Fig. 11. The region of the trigeminal ganglion, the boundary cap, and the trigeminal nerve in a day E16 embryo. Note that the peripheral branch of the trigeminal nerve contains many cells (presumably Schwann cell precursors) that are contiguous with the boundary cap, but the central portion of the nerve is devoid of cell bodies. Methacrylate. *Scale:* 100 μm

derivation), is the source of Schwann cells of the trigeminal ganglion and nerve. Distal fibers of the trigeminal ganglion were first seen, in small numbers and in the vicinity of the ganglion, in day E13 embryos (Fig. 10A). By day E15, the ophthalmic (Fig. 12A) and the maxillary and mandibular nerves (Fig. 12B) have approximated their target structure.

Comments. Our results support Hamburger's (1961) conclusion that the two lobes of the trigeminal ganglion are not derived from different germinal sources. Hamburger found that after destruction of the medulla in the chick embryo, the shared placodal epidermis could by itself assure the growth of all three branches of the trigeminal nerve. However, our results do not support his conclusion that the small neurons of the ganglion originate from the neural crest and the large neurons from the epidermal placode. In agreement with observations in chicks (e.g., Hamburger 1961; Johnston 1966; Noden 1975), we could distinguish two cell aggregates in the trigeminal anlage of E11 and E12 rat embryos. In day E13 embryos the cells of the lateral aggregate, which could well be of placodal origin, were distinctly larger than the cells of the medial aggregate, which appeared to be of neural crest derivation. However, by tracing the fate of cells in these two regions in closely spaced embryos from day E14 onward, we came to the conclusion that the lateral cell aggregate is the source of all trigeminal neurons, both large and small, while the medial cell aggregate produces a transient structure, the trigeminal boundary cap.

Cells in the lateral trigeminal anlage began to increase in size by day E13, the day after 20% of the neurons were produced (Fig. 5). By day E14, the two separate lobes of the ganglion could be distinguished at this site, with young differentiating neurons

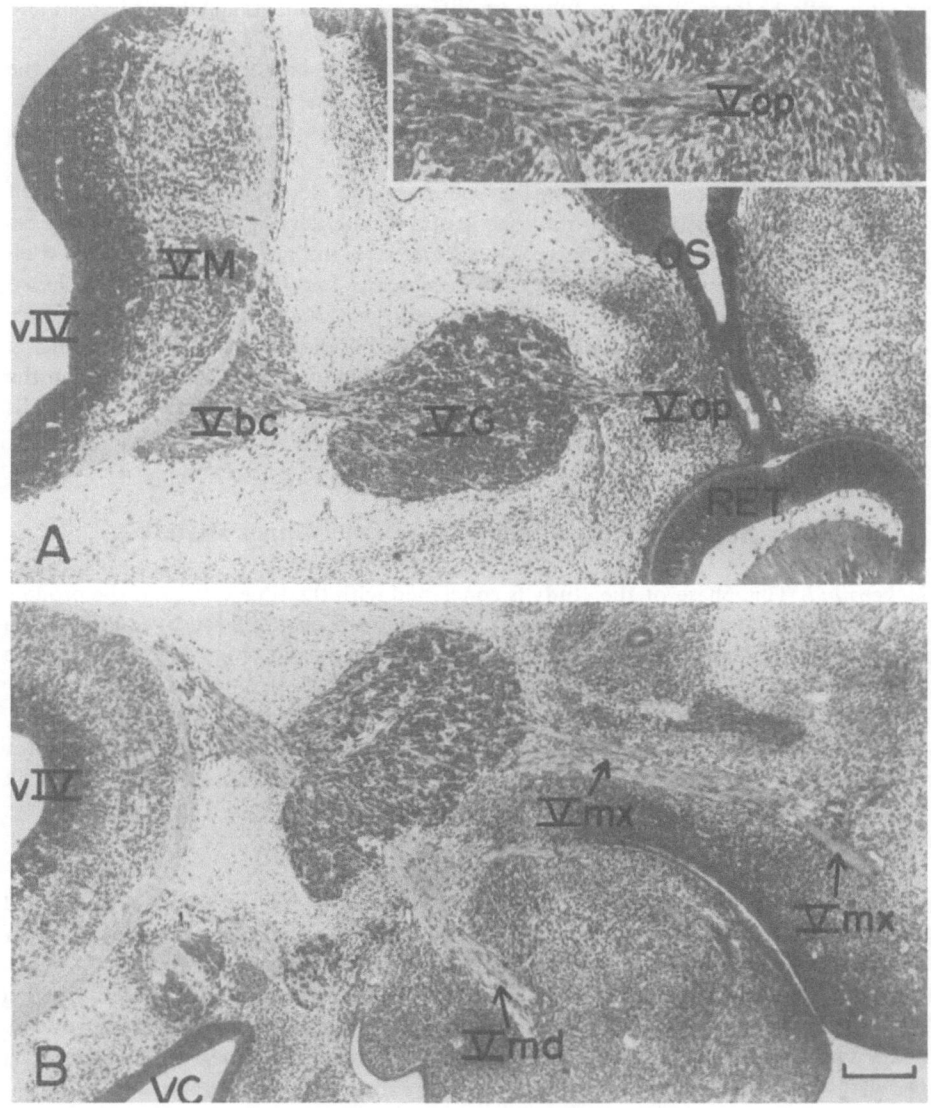

Fig. 12. *A* Horizontal section from a day E15 embryo showing the ophthalmic nerve in the vicinity of the eye (enlarged in the *inset, upper right corner*). *B* Sagittal section from a day E15 embryo with the maxillary and mandibular nerves. Methacrylate. *Scale:* 200 μm

scattered throughout the structure. By day E15, at a time when neuron production was still in progress, a range of neurons of different sizes was evident without any obvious evidence of regional segregation. The apparent random scattering of large and small neurons persisted on the subsequent days, while satellite cells began to appear in increasing numbers. Our observations failed to provide support for the report, made in the chick (Gaik and Farbman 1973b), that at an early stage of development the small

15

ganglion cells become dark. We found no differences in the stainability of small and large cells up to day P5 (Fig. 9).

In contrast to the cells produced in the region of the lateral trigeminal anlage, the cells derived medially remained small throughout the entire period of prenatal development. The boundary cap formed by these cells was traversed by trigeminal afferents as early as day E13, and by trigeminal efferents on day E14. From day E15 onward, strings of cells could be traced from the boundary cap along the proximal portion of the trigeminal nerve peripherally, but the central (medullary) portion of the nerve remained free of cells for several days. The trigeminal boundary cap was still present on day E22 embryos, but could no longer be recognized in a P5 rat pup. We concluded that the trigeminal boundary cap, which may be of neural crest derivation, is the source of peripheral Schwann cells; the cap region presumably disappears during the perinatal period. We cannot at present resolve the discrepancy between our conclusions and the claims made in the chick, on the basis of experimental work, that the neural crest cells are the source of small neurons of the trigeminal ganglion.

3.3 Development of the Trigeminal Nuclei of the Upper Medulla

Background. This phase of the study is concerned with the time course of the production of neurons of the trigeminal motor nucleus and the principal sensory nucleus, and the growth of trigeminal afferents and efferents. The purpose is to attempt to correlate these events centrally with concurrent processes in the trigeminal system peripherally. This embryonic investigation was done with the aid of our previous thymidine-radiographic datings (Altman and Bayer 1980d), which showed that neurons of the motor nucleus of the trigeminal nerve are produced between days E11 and E12 (with a peak on day E12), and neurons of the principal trigeminal sensory nucleus between days E13 and E16 (with a peak on day E14).

Results. Because of the cephalic flexure, the orientation of the upper medulla and pons in embryos is different from that in adults. Figure 13 shows the orientation of the relevant structures on day E15 in sagittal and horizontal sections, to help the visualization of the spatial relations of germinal cell sources, migrating cells, and growing fibers in the early embryo. The trigeminal ganglion, which occupies a ventral position in the adult, is in a rostral position in the embryo, and what appears to be the rostral medulla is really its ventral aspect with the basal plate neuroepithelium.

Spindle-shaped cells, apparently in the process of migration from the basal plate of the upper medulla in the direction of the trigeminal boundary cap, were recognized in day E13 rats (Fig. 14). The settling of motor neurons laterally in the vicinity of the boundary cap, while cell migration was still in progress, was evident by day E14 (Figs. 15, 16A). Also on day E14, a few efferent fibers of the settling trigeminal motor neurons began to leave the medulla, traversing a component of the trigeminal boundary cap (Fig. 15B). The settling of the motor neurons was continuing on day E15. By day E16, spindle-shaped cells were no longer seen in this region.

A stream of spindle-shaped cells could be traced in day E15 rats (Fig. 16B), in apparent migration from the region of the alar plate of rhombomere 1 toward the trigeminal boundary cap. We identified these cells as the neurons of the principal nucleus, which are produced with a peak on day E14 (Altman and Bayer 1980d). A comparison

16

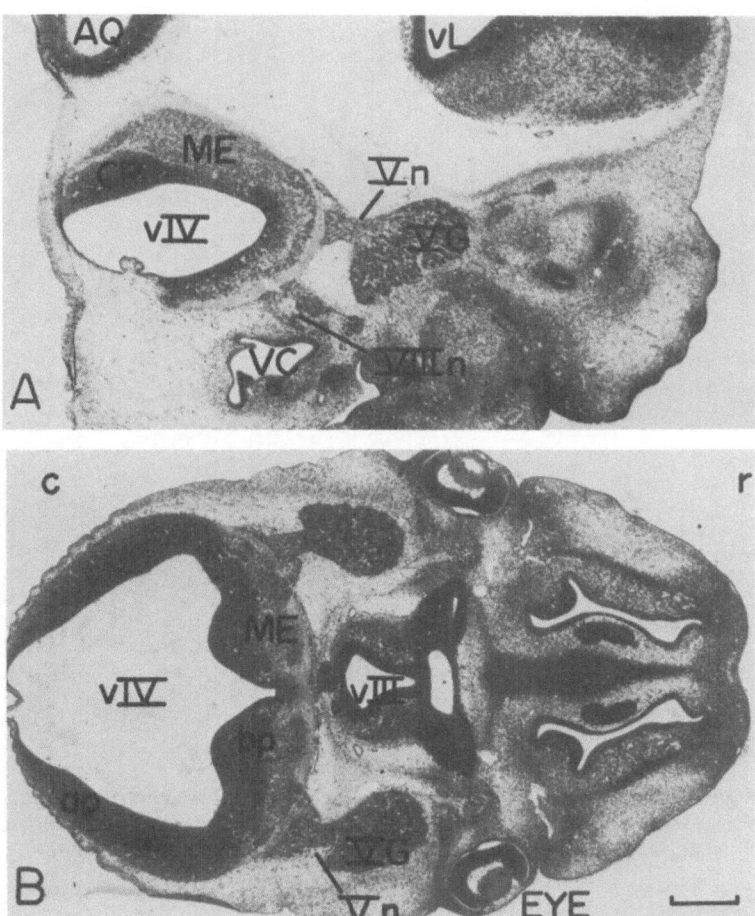

Fig. 13. Spatial relations of the trigeminal ganglion, upper medulla, and future pons with the face in day E15 embryos in sagittal *(A)* and horizontal *(B)* sections. Methacrylate. *Scale:* 500 µm

of this region of the upper medulla in day E14 (Fig. 16A) and day E15 (Fig. 16B) rats indicates that, due to this second wave of cell migration in the direction of the trigeminal boundary cap, the neurons of the trigeminal motor nucleus are displaced medially by the later-arriving neurons of the principal nucleus. The settling of neurons of the principal nucleus appeared completed on day E17, and by this time the position of the motor nucleus and principal sensory nucleus resembled the mature pattern.

Comments. According to our thymidine-radiographic datings (Altman and Bayer 1980d), the production of trigeminal motor neurons peaks and ends on day E12. On the next day, in day E13 rats, we could trace a stream of migrating cells from the medial aspect of rhombomere 1 in a lateral direction. The onset of migration in the direction of the incipient boundary cap coincided with the arrival of the first trigeminal afferents in the same region peripherally. By day E14 the first contingent of motor

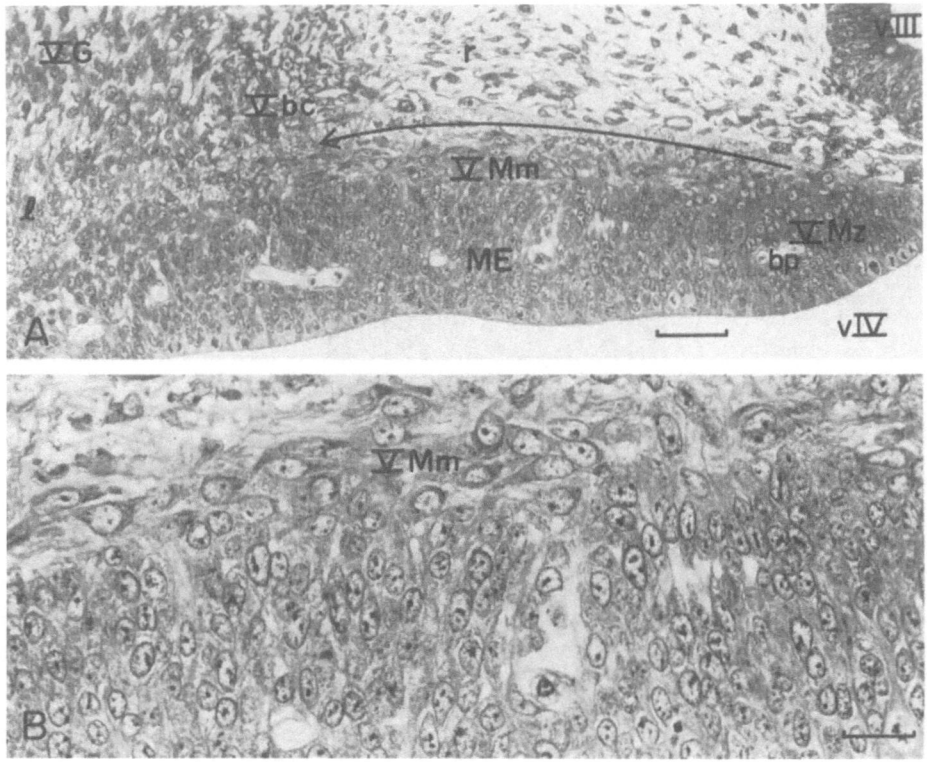

Fig. 14. *A* Horizontal section from a day E13 embryo with the apparently migrating cells of the trigeminal motor nucleus in mid-position between their germinal source and settling site. *B* Higher magnification of the spindle-shaped young neurons. Methacrylate. *Scales: A*, 50 μm; *B*, 20 μm

neurons were appearing near the border of the medulla in the vicinity of the boundary cap, and their settling appeared completed by day E15. At about this time, a new wave of migrating cells began to move in this direction from the lateral flank of rhombomere 1. These were identified as the neurons of the principal sensory nucleus, which are generated with a peak on day E14. Thus in the case of both nuclei, cell migration occurred 1 day after peak level of cell generation and both cell groups moved toward the trigeminal boundary cap. The spatiotemporal relations of cell production and migration, and the growth of afferents and efferents, are summarized in Fig. 17. It is tempting to speculate that the trigeminal boundary cap serves as a guidepost and is involved in the assembly of the entire lower trigeminal system derived from different germinal sources (the trigeminal placode, and the basal plate and alar plate of rhombomere 1). This proposition will be examined later in the light of observations in the other cranial nerve ganglia and nuclei.

A comparison of the temporal pattern of neuron production obtained in this study with our datings of other components of the trigeminal system (Altman and Bayer 1980. Fig. 16) suggests that in this system the motor neurons are produced first (peak on day E12), the primary sensory neurons next (peak on day E13), and the second order sensory neurons of the upper and lower medulla last (with peaks ranging

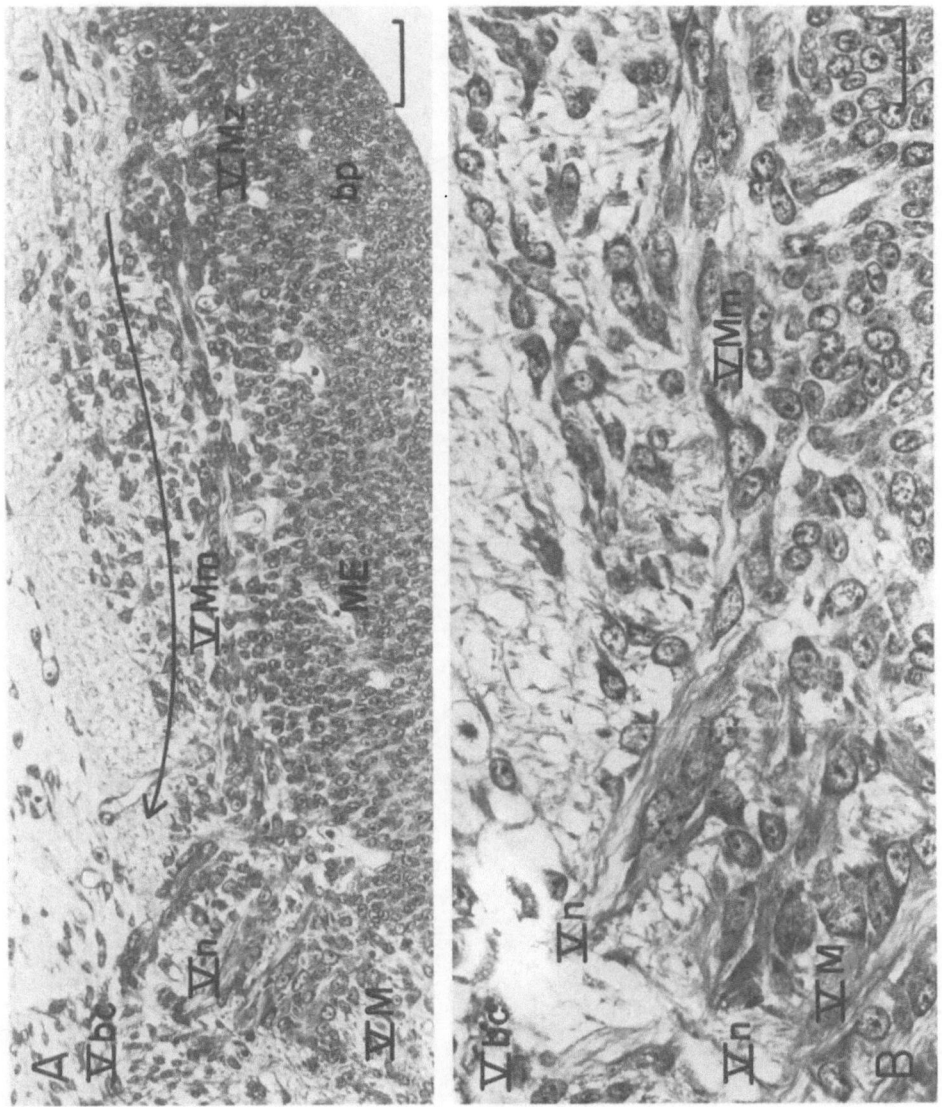

Fig. 15. *A* Horizontal section from a day E14 embryo with migrating cells and settling trigeminal motor neurons. *B* The exiting efferents of the motor nucleus shown at higher magnification. Methacrylate. *Scales: A*, 50 μm; *B*, 20 μm

from E14 to E15). The pattern of production of neurons of the mesencephalic nucleus of the trigeminal (with over 80% of the cells produced on day E11 or earlier; Altman and Bayer 1980d) does not match that of neurons of the trigeminal ganglion, although a shared embryonic derivation has been proposed (Piatt 1945; Narayanan and Narayanan 1978). It may be noted in this context that, in spite of the earlier production of trigeminal motor neurons than ganglion cells, the trigeminal efferents do not start to

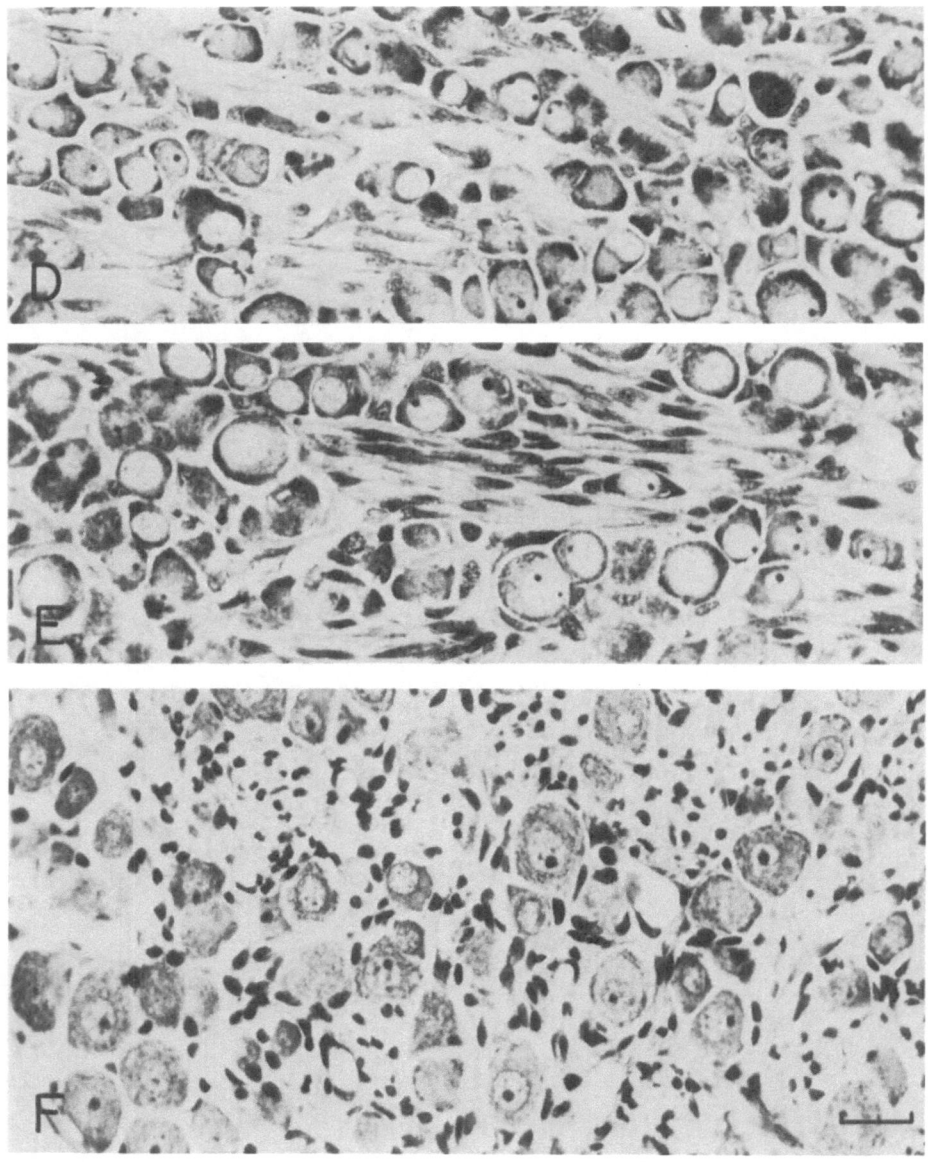

Fig. 16. *A* Horizontal section from a day E14 rat with the accumulating cells of the trigeminal motor nucleus laterally near the boundary cap. *Horizontal arrow* is over the migrating cells of the motor nucleus (compare with Fig. 15). *Small vertical arrow* points to the exiting efferents of the motor nucleus. *B* A matched section from a day E15 rat, showing spindle-shaped cells apparently moving toward the boundary cap from the alar plate *(oblique arrow)*. As a result of this second wave of migration the trigeminal motor neurons are displaced medially. Methacrylate. *Scale:* 100 μm

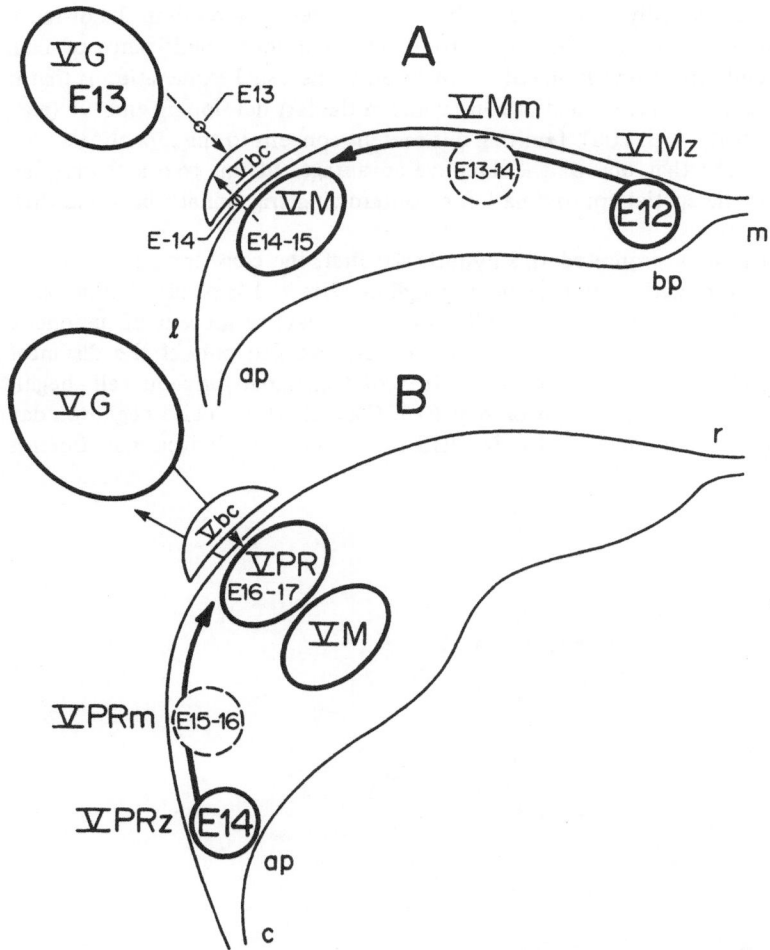

Fig. 17 A, B. Schematic illustration of two stages in the development of the lower trigeminal system. *A* The trigeminal motor neurons are mostly produced on day E12 in the medullary basal plate medially, migrate on the following days, and settle laterally near the trigeminal boundary cap on days E14–E15. The boundary cap is reached by afferents of the trigeminal ganglion on day E13 and by trigeminal efferents on day E14. *B* The neurons of the principal sensory nucleus are produced in the medullary alar plate with a peak on day E14, migrate on the following days rostrally, and settle in the vicinity of the boundary cap on days E16–E17, displacing the motor nucleus medially

grow before the trigeminal afferents, presumably because the earlier-produced motor neurons first have to migrate to the boundary region of the medulla.

Following neurulation, a prominent feature of embryonic development in vertebrates is the formation of flexures in the body axis. In the rat, the straight body axis begins to flex anteriorly and posteriorly early on day E11, at about the time when the first somites become recognizable. The cephalic flexure is completed by day E12, and it results in bringing the face region in contact with the ventral aspect of the upper trunk. The cephalic flexure remains prominent through day E15 (Fig. 13), but thereafter the flexion is gradually reversed and there is a progressive growth in the distance between the eye and jaws, and the upper trunk. The functional significance of axial flexions during embryonic development is not known; the usual explanation is that it is a mechanical consequence of spatial constraints in the fast-developing embryo (e.g., Patten and Carlson 1974; p. 200). Limiting our present concern to the cephalic flexion, we propose (Fig. 18) that one of the adaptive advantages of this transient morphogenetic process is the spatial approximation of outgrowing trigeminal fibers and their target structures.

Two considerations prompted this hypothesis: first, the close temporal relationship between the production of trigeminal ganglion cells and cephalic flexion; and, second, evidence from tissue culture studies that the selective growth of peripheral (autonomic) fibers toward their target structures is successful only if the distances are relatively small. As we saw, the production of trigeminal ganglion cells begins slowly on day E11 and is completed on day E15. Cephalic flexion also begins on day E11 and is completed on day E12. By day E13 the outgrowth of trigeminal afferents

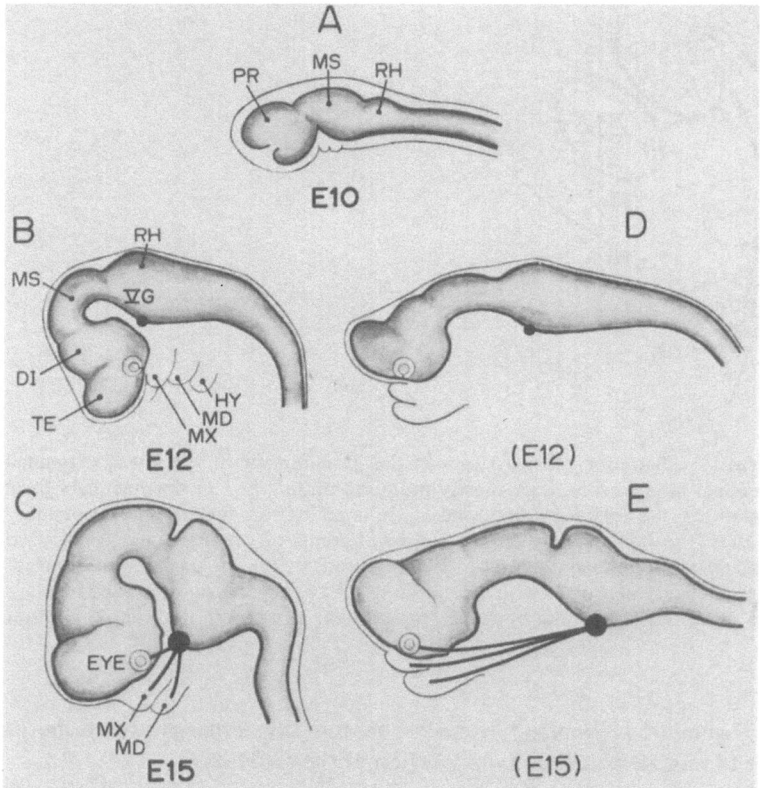

Fig. 18. *A* Rostral portion of the neural tube of the day E10 rat embryo prior to cephalic flexion. *B* Completion of cephalic flexure by day E12, at about the time of onset of neuronal differentiation in the trigeminal ganglion. *C* Outgrowth of distal trigeminal afferents in large numbers in the day E15 embryo at the time when the production of ganglion cells comes to an end. Based on various classical sources and adapted to the rat. *D* and *E* show the large distances that trigeminal afferents would have to traverse in the absence of cephalic flexion. Drawings are not to scale

is evident both proximally and distally. It is not unreasonable to assume that the polarization of ganglion cells and the initial steps of axonal growth begin immediately after differentiation on day E12, and that the now neighboring eye and jaw tissue primordia exert a guiding influence on the growth of the earliest distal afferents. By day E15, the bulk of the trigeminal nerve has grown appreciably, and its three branches are recognizable near the structures they innervate. Throughout this period the distance from the border of the trigeminal ganglion and its target structures did not exceed 0.5–1.0 mm. Recent in vitro studies have shown that sympathetic fibers in a culture medium grow preferentially toward their target structures if these are placed no farther than 1–2 mm away (Chamley et al. 1973; Chamley and Dowel 1975). Similar effects have been obtained with parasympathetic ganglia, which show preferential neurite growth toward their targets placed at distances of up to 0.5 mm (Coughlin 1975). Accordingly, we postulate that cephalic flexion is a morphogenetic adaptation that allows the effective diffusion of some chemotactic guidance agent from the target structures of the head at a time when the fibers innervating them begin to grow. Such a morphogenetic adaptation is not necessary in all parts of the body; for instance, the motor axons in the chick have to grow less than 1.0 mm to reach the limb muscles (Jacobson 1978, p. 158). However, flexion in other regions of the body, such as the tail, may find a similar explanation.

In summary, our thymidine-radiographic datings and embryonic observations suggest a synchronization of several apparently independent developmental events in the morphogenesis of the lower trigeminal system. These events include the precise timing of neuron production in central and peripheral germinal matrices; the convergent migration of sensory and motor neuron from two separate germinal sources; the timed outgrowth of efferents and ingrowth of proximal afferents at the same site in the medulla (possibly guided by a transient cellular aggregate, the boundary cap); and finally, the apparent bending of the entire head region in the direction of the growing trigeminal afferents and efferents.

4 Development of the Facial Ganglion in Relation to the Facial Motor Nucleus

4.1 Time of Origin of Facial Ganglion Cells

Background. The facial (geniculate) ganglion contains the perikarya of the sensory neurons of the seventh cranial nerve. The distal (peripheral) axons of these neurons form the nervus intermedius and contribute some afferents to the predominantly efferent seventh nerve (Foley and Dubois 1943; Bruesch 1944; Buskirk 1945). The nervus intermedius has two branches: the greater superficial petrosal nerve, which reaches the sphenopalatine ganglion; and the chorda tympani, which joins the lingual nerve and is distributed over the frontal portion of the tongue. The afferents of the facial nerve proper reach the external ear (Brodal 1981). This multiple distribution of facial afferents is supported by a physiological study (Boudreau et al. 1971) which showed that different geniculate ganglion cells in the cat responded selectively to displacement of hairs of the ear, mechanical stimulation of the soft palate and pharynx, and chemical or mechanical stimulation of the tongue.

In general, the proximal (central) afferents of the seventh nerve have little or no direct relationship to the large motor nucleus of the facial nerve, but distribute to two sensory nuclear complexes, the solitary nucleus and the spinal nuclei of the trigeminal nerve (Rhoton 1968; Beckstead and Norgren 1979; Nomura and Mizuno 1981). The fibers terminating in the rostral portion of the solitary nucleus are generally held to mediate gustatory sensations; the possible functions of the fibers terminating in the trigeminal nuclei are less clear. Although physiological studies support a projection of facial afferents to the trigeminal nuclei (Iwata et al. 1972; Tanaka 1977), the idea that the facial afferents mediate facial reflexes (Iwata et al. 1972) is not supported by the available studies. Rather, it appears that the afferents of the facial motor nucleus that mediate facial reflexes of the eye and vibrissae come from the trigeminal nerve or nuclei. For instance, it has been found that the blink reflex, which is executed by facial efferents, is abolished by severance of the trigeminal nerve (Kugelberg 1972), by lesions of the principal nucleus of the trigeminal (Tokunaga et al. 1958), and by trigeminal tractotomy (Lindquist 1973). Thus, there is a complex relationship between the facial and trigeminal ganglia and nuclei, in that the facial reflex pathways that have been analyzed are not composed of facial afferents reaching the trigeminal nuclei, but of trigeminal afferents. The link that this necessitates from the trigeminal nuclei (particularly their spinal components) to the facial motor nucleus has repeatedly been demonstrated anatomically (Carpenter and Hanna 1961; Stewart and King 1963; Dom et al. 1973; Burton et al. 1979; Erzurumlu and Killackey 1979) and also physiologically (Tanaka et al. 1971).

As far as we know, the time of origin of neurons of the mammalian facial ganglion has not previously been described. In this phase of the study we sought to determine the birth dates of facial ganglion cells in relation to the other cranial nerve ganglia peripherally, and centrally with reference to selected nuclei of the medulla, particularly the facial motor nucleus.

Results. In the rat, the facial ganglion is a spherical structure (Fig. 19), by far the smallest of all the cranial nerve ganglia. It is nestled in a recess of the temporal bone anterodorsal to the cochlea, immediately behind the caudal tip of the trigeminal ganglion. The neurons of the facial ganglion range in size from small to large (Fig. 20B) and are similar in appearance to the neurons of the trigeminal ganglion (Fig. 20C). However, in animals injected with ^{3}H-thymidine at different ages, the labeling patterns were different in the two ganglia (Fig. 20B, C).

In the offspring of dams injected on days E11+12 all facial ganglion cells were labeled, with labeling intensity ranging from light to heavy. A small proportion of the neurons (10%) were no longer labeled in the E12+13 injection group, and many (62%) were no longer labeled in the E13+14 group. In the E14+15 group only a few scattered neurons (4%) were labeled and these were typically small cells. No neurons were found with a label in the E15+16 group. However, all the satellite cells were labeled in the latter group, and in all the animals available to us with later injections (up to days E17+18).

The quantitative results are summarized in Fig. 21. The neurons of the facial ganglion are produced between days E11 and E14, with peak production time on day E12 (52%) and a large complement acquired on day E13 (34%).

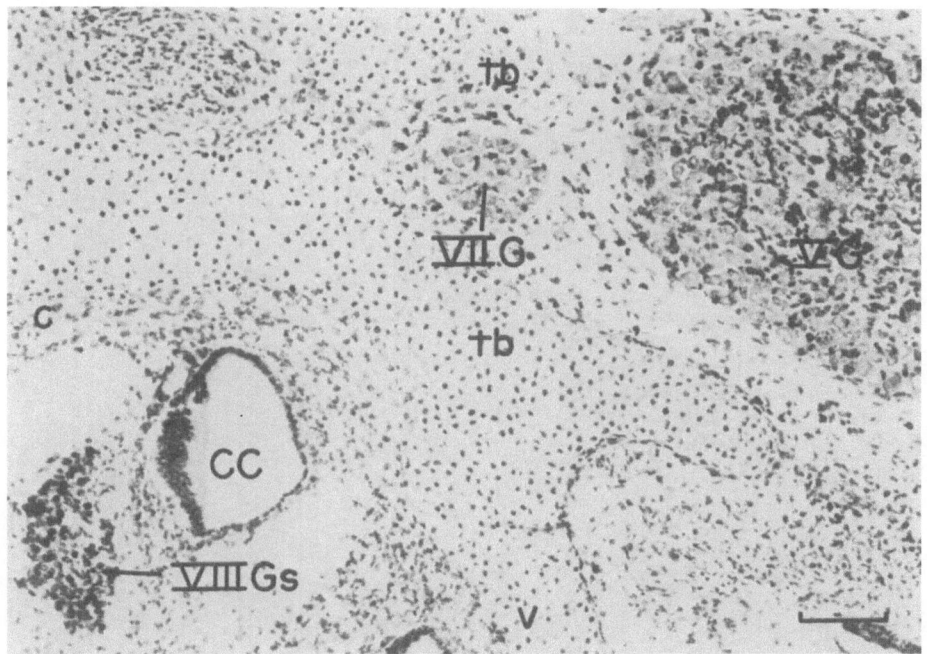

Fig. 19. The small facial ganglion in the recess of the temporal bone near the anterior surface of the cochlea in sagittal section (compare with Fig. 1). All neurons of the spiral ganglion, some neurons of the trigeminal ganglion, but none of the facial ganglion (except its satellite cells) are labeled. Injections on days E14+15. Paraffin. *Scale:* 100 μm

Comments. These results indicate that neurogenesis in the facial ganglion (peak on day E12) precedes the production of neurons in the trigeminal ganglion (peak on day E13; Fig. 5). Surprisingly, this facial-to-trigeminal cytogenetic gradient in the production of primary sensory neurons is not matched centrally in terms of the temporal order of production of facial and trigeminal motor neurons. According to our recent studies, the trigeminal motor neurons are produced between days E11 and E12, with a peak on day E12 (Altman and Bayer 1980d), and the facial motor neurons between days E12 and E14, with a peak on day E13 (Altman and Bayer 1980b), resulting in a trigeminal-to-facial cytogenetic gradient. Looking at these results (Fig. 22) from a different point of view, it appears that in the trigeminal system the central motor neurons are produced before the peripheral sensory neurons, whereas in the facial system the peripheral sensory neurons precede the motor neurons. The generally held view that, at corresponding levels of the neuraxis, the motor neurons are generated before the sensory neurons seemed to obtain support in the preceding study of the trigeminal system, but is contradicted by our present results.

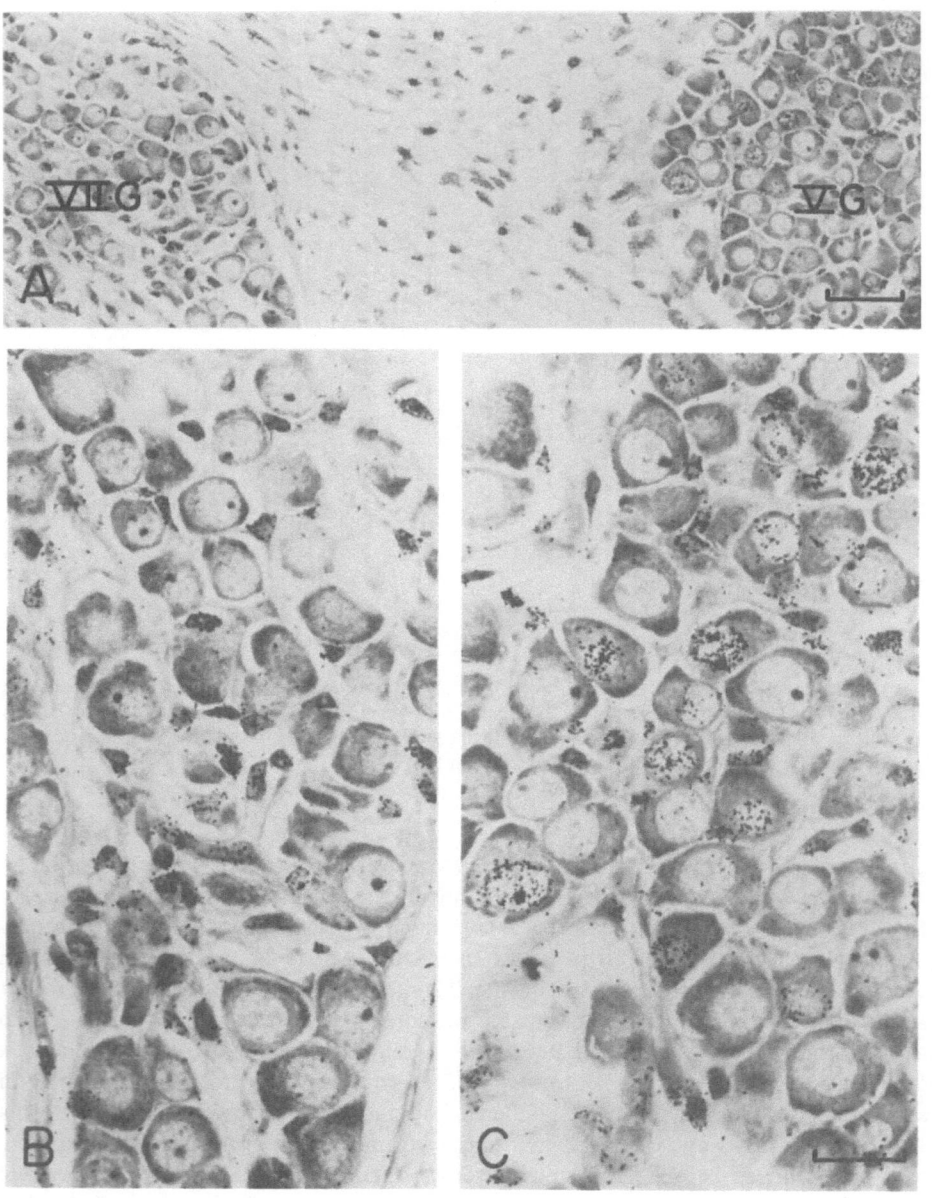

Fig. 20. *A* The facial and trigeminal ganglia in an embryo labeled on days E14+15. *B* The facial ganglion and *C* the trigeminal ganglion at higher magnification. Many trigeminal neurons are labeled, but in the facial ganglion only satellite cells are labeled. Methacrylate. *Scales: A*, 50 μm, *B, C*, 20 μm

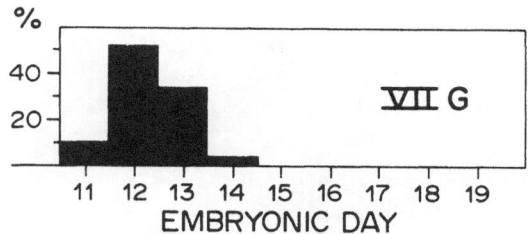

Fig. 21. Percentage of facial ganglion cells produced between days E11 and E14

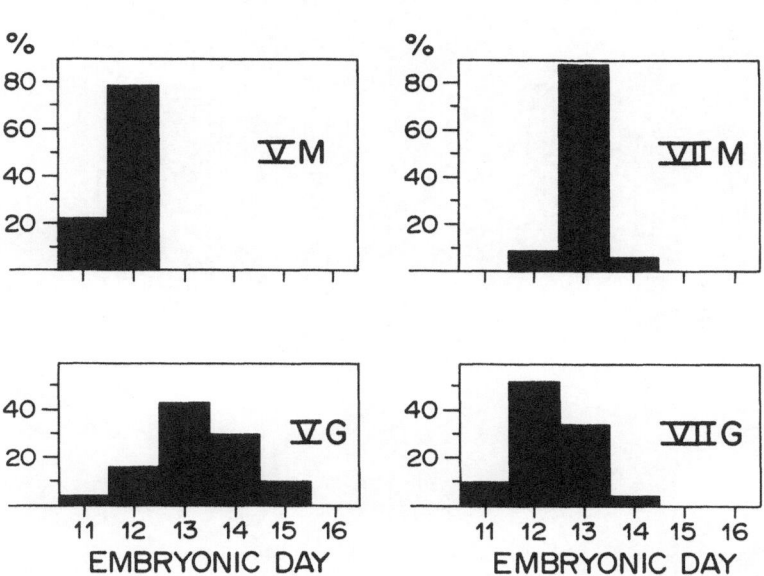

Fig. 22. The reversed relationship between the order of production of motor neurons and primary sensory neurons (ganglion cells) in the trigeminal system *(left panel)* and the facial system *(right panel)*. Sources referenced in the text

4.2 Embryonic Development of the Facial Ganglion

Background. The embryonic development of the facial ganglion has received some attention in the past, in particular with reference to the presumed derivation of the facial ganglion from a shared acousticofacial primordium, as originally proposed by Weigner (1901). In contrast, Streeter (1906) maintained that a separate facial anlage was identifiable from the earliest stages of embryonic development in man. The latter view was supported by several investigators working with amphibia (reviewed by Yntema 1937), and by Batten (1958), who studied the development of sheep. Nevertheless, the idea that the facial and otic ganglia originate from a shared germinal source appeared to prevail (e.g., Kimmel 1941, in the rat), and it appears as an unqualified assertion in most textbooks of embryology (e.g., Hamilton et al. 1964, p. 365; Patten and Carlson 1974, p. 416). There has been a similar debate as to whether the facial ganglion is derived from the neural crest or placode. Yntema (1937) and Batten (1958) subscribed to the strictly placodal origin of the facial ganglion, but most textbooks state (e.g., Hamilton et al. 1964, p. 365) that the geniculate ganglion is derived partly from

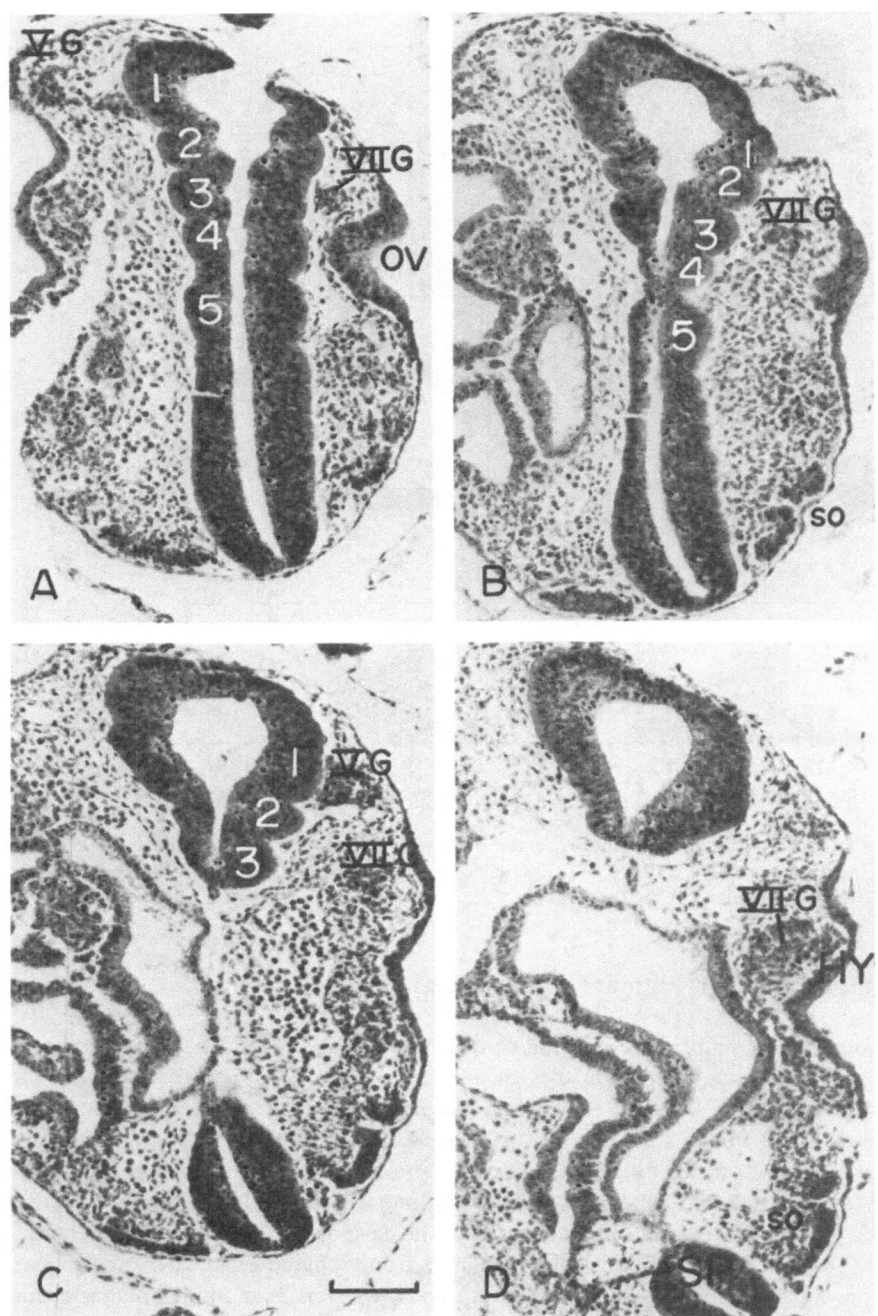

Fig. 23 A–D. Horizontal sections from a day E11, two-somite rat showing the anlage of the facial ganglion in relation to the rhombomeres *(numerals)* and the placode of the hyoid arch. Paraffin. *Scale:* 100 μm

the neural crest and partly from the ectodermal placode of the second (hyoid) branchial arch.

Results. The anlage of the facial ganglion is recognizable in day E11 rats with two to three somites before the closure of the otic vesicle (Fig. 23). The facial rudiment extends from the vicinity of rhombomere 3 medially and dorsally (Fig. 23A, B) to a cell condensation of the hyoid arch laterally and caudally (Fig. 23C, D). In day E12 rats (Fig. 24), the placode in close contiguity with the facial rudiment was recognizable in the cleft region between the hyoid arch and maxillomandibular enlargement (Fig. 24C), the lateral aspect of the mouth cavity. At this age a small complement of cells emerged from the rostral aspect of the otic vesicle; as we shall describe in detail in the next section, this is the rudiment of the otic ganglion the neurons of which are generated later than the neurons of the facial ganglion. The otic ganglion was contiguous with the facial ganglion dorsally (Fig. 24A), but the two were separated more caudally (Fig. 24B), where contiguity with the hyoid arch region was maintained (Fig. 24C). By day E13 (possibly as early as day E12; Fig. 36) the facial ganglion assumed its spherical shape (Fig. 25) and its neurons were beginning to show signs of cytological differentiation (Fig. 25B).

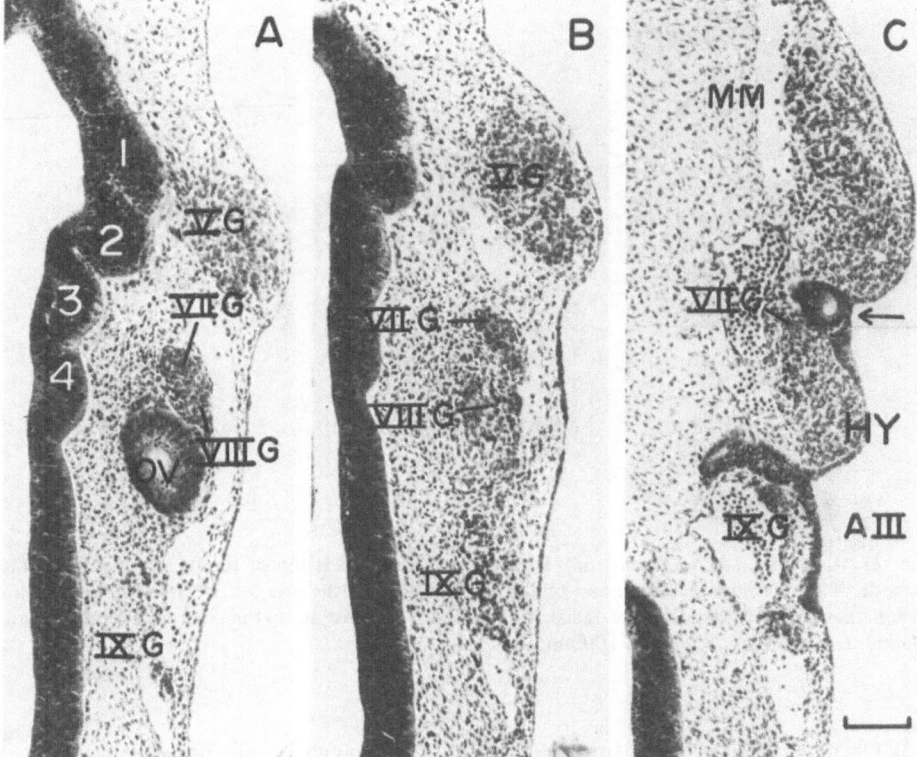

Fig. 24 A–C. Horizontal sections from a day E12 rat to show the contiguous but separate facial and otic ganglia *(A* and *B)*. *Arrow* in *C* points to the cleft region of the maxillomandibular enlargement and the hyoid arch. *Numerals* refer to the rhombomeres. Paraffin. *Scale:* 100 μm

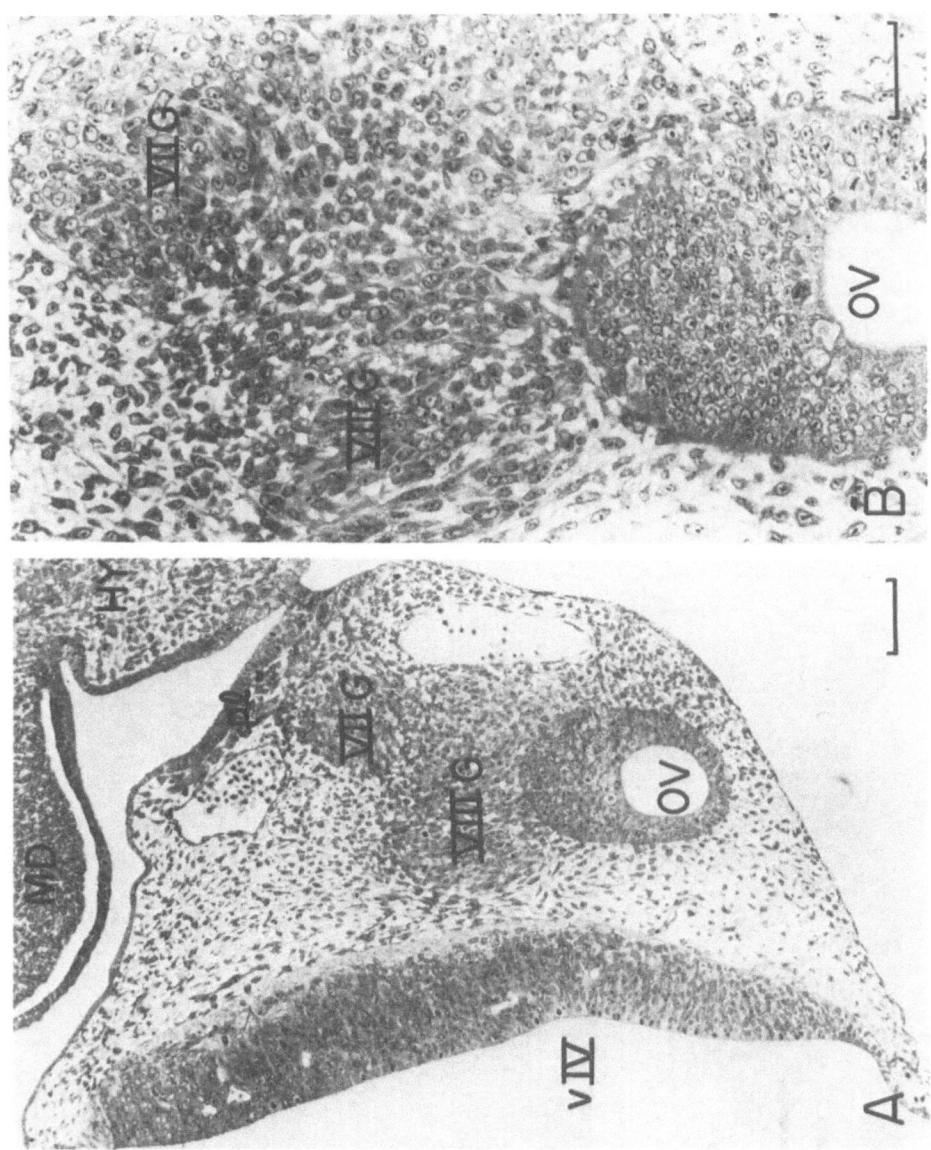

Fig. 25. *A* Oblique section from a day E13 rat to show the relation of the facial ganglion to the placode of the hyoid arch region and of the otic ganglion to the otic vesicle. *B* At higher magnification the spheroid shape of the facial ganglion and onset of its cytological differentiation are evident. Methacrylate. *Scales: A*, 100 μm; *B*, 50 μm

On day E14, by which time 96% of the facial ganglion cells have been generated (Fig. 21), the morphological differentiation of these neurons far surpassed the differentiation of the otic ganglion cells (Fig. 26) and the facial neurons assumed a bipolar shape with densely staining cytoplasm at the two poles (Fig. 27). The early maturation

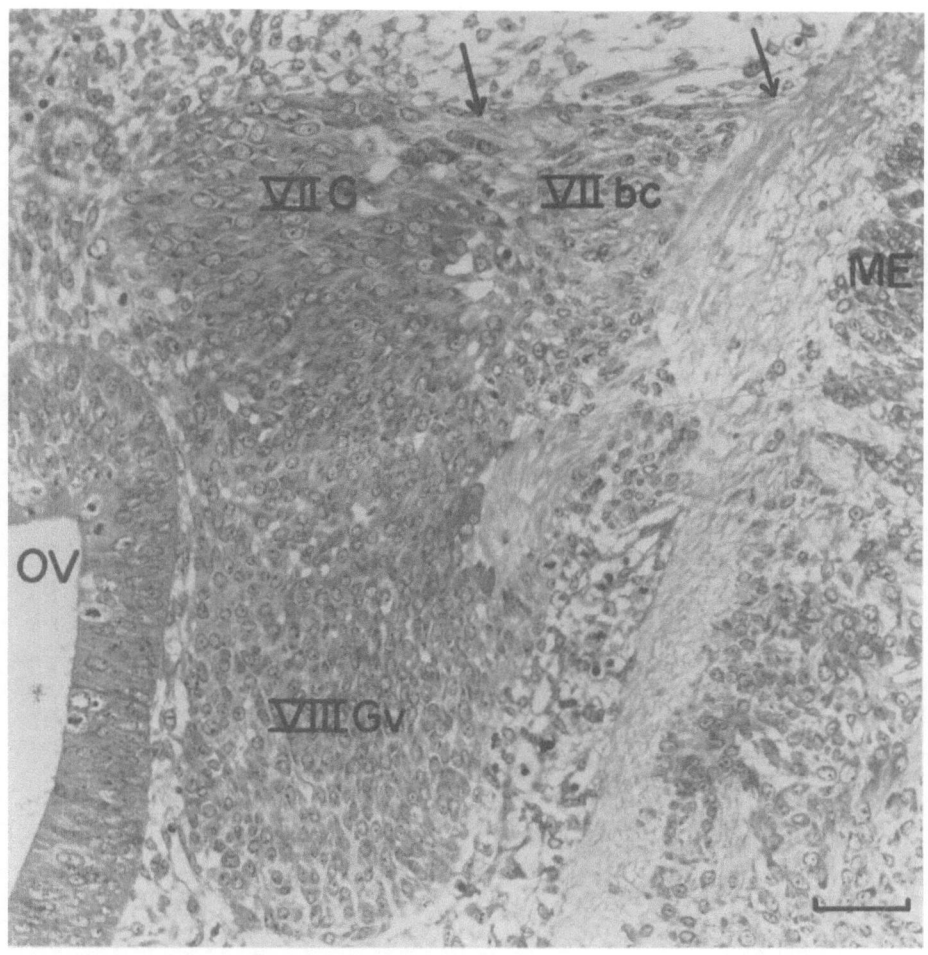

Fig. 26. Horizontal section through the region of the facial and vestibular ganglia from a day E14 rat. Advanced maturation of the facial ganglion cells with respect to the vestibular ganglion cells is evident. However, the cells of the facial boundary cap remain primitive. *Arrows* point to the presumed proximal afferents of the facial ganglion. Methacrylate. *Scale:* 50 μm

of the early-generated facial ganglion cells was also evident in comparisons with the neurons of the trigeminal ganglion between days E14 and E17, but the difference was no longer evident by day E21 (Fig. 20).

Comments. The thymidine-radiographic evidence of the early production of facial ganglion cells was supported by embryological observations. Although the facial ganglion is a much smaller structure than the trigeminal ganglion, its anlage matched in size the trigeminal primordium in day E11 rats. Also, the cytological differentiation of facial ganglion cells preceded that of the trigeminal ganglion cells, and this lead was evident for several days after the onset of differentiation. We have referred earlier to statements that the facial ganglion is derived, in common with the eighth nerve ganglia,

31

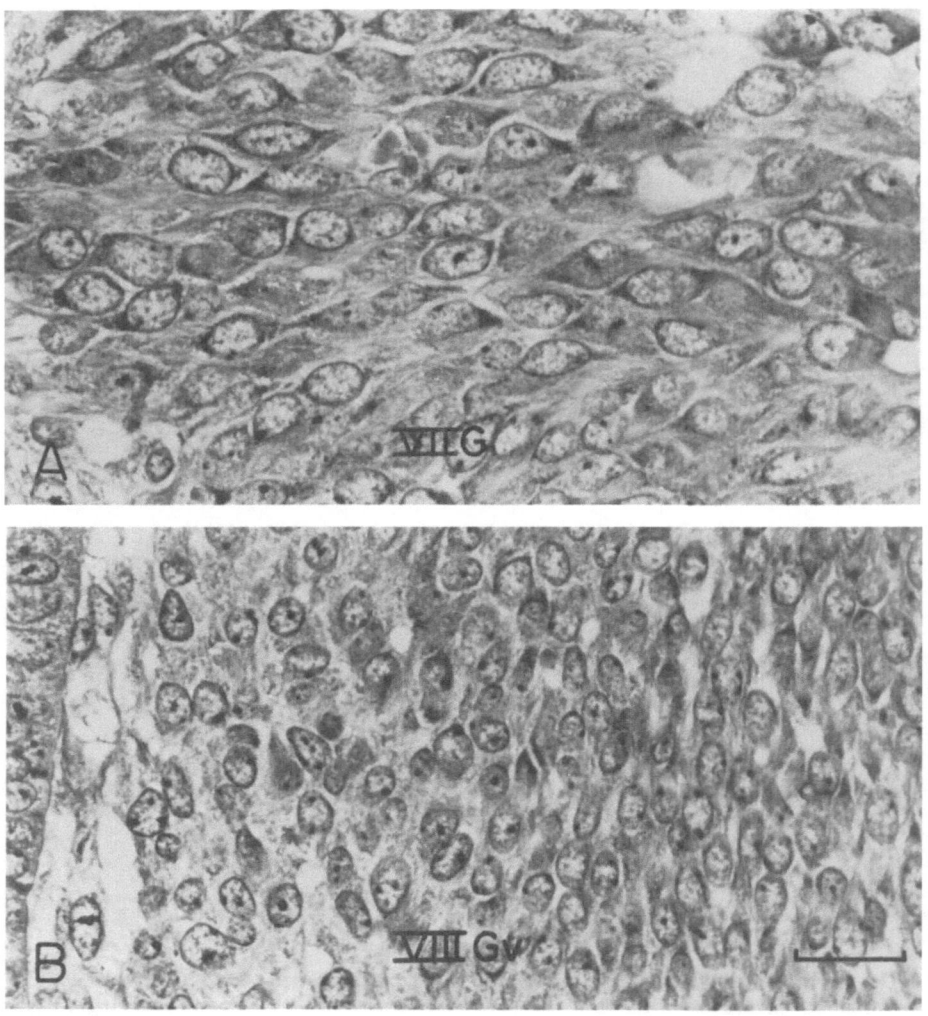

Fig. 27. Higher magnification of cells of the facial ganglion *(A)* and vestibular ganglion *(B)* illustrated in Fig. 26. Methacrylate. *Scale:* 20 μm

from a shared acousticofacial primordium. But our observations support Batten's (1958) conclusions that the two originate from different primordia. The anlage of the facial ganglion was recognizable as a discrete structure before the anlage of the otic ganglion appeared, and the maturation of the facial neurons antedated the maturation of the vestibular ganglion cells. Moreover, our observations indicated that the facial ganglion derived from a superficial ectodermal thickening (placode) in the cleft region between the hyoid arch and the maxillomandibular thickening, and the otic ganglion, as we shall see later, derives from the ectoderm of the otic vesicle.

4.3 Development of the Facial Motor Nucleus and Nerve

Background. The anomalous central course of facial efferents is well known. Instead of leaving the medulla directly from the ventral position where the perikarya of the facial nucleus are located, the facial nerve begins to course in the opposite direction dorsomedially, loops around the abducent nucleus, traverses the medulla in a ventrolateral and rostral direction, and finally exits. Comparative studies (reviewed in Ariëns Kappers et al. 1936, Vol 1) have shown that in many lower vertebrates the facial nucleus is dorsally situated, while in others some parts or all of the facial nucleus shift ventrally and caudally. Ariëns Kappers proposed that the facial motor neurons in mammals migrate from their site of production dorsally, under the influence of neurobiotaxis (Ariëns Kappers 1917), in the direction of gustatory fibers. In this study we attempted to trace the time course and sequence of events in the migration of the facial motor neurons and the growth of the facial nerve in rat embryos.

Results. According to our datings (Altman and Bayer 1980b: Fig. 2), peak production of facial motor neurons occurs on day E13, with a rostral-to-caudal gradient. In day E14 rats (Fig. 28), we identified the facial motor nerve in horizontal sections as a broad band of fibers coursing from the midline region (the alar plate) laterally in the direction of the facial ganglion (Fig. 28A, B). The fibers left the medulla in several fascicles and traversed an aggregate of cells which we consider to be the facial boundary cap (Figs. 28C, 29) before reaching the facial ganglion. The course of the facial nerve in day E14 rats in coronal section is shown in Fig. 29. Again the nerve can be seen coursing from the alar plate over the surface of the neuroepithelium laterally; then, reaching the region of the facial boundary cap, it leaves the medulla in several fascicles. In coronal sections the perikarya of the facial efferents can be recognized as differentiating spindle-shaped cells situated in the vicinity of the midline neuroepithelium (Fig. 30). Peripherally, the facial efferents could be traced to the facial ganglion where they joined the afferents of the facial nerve (Fog. 31B). The facial afferents, like the efferents, traverse the facial boundary cap but enter the medulla more laterally, in the region of the alar plate (Fig. 31A).

On day E14, the facial nerve follows a relatively straight lateral course from the midline. The motor neurons begin to migrate caudally and ventrolaterally on day E15 (Fig. 32A, B), and some reach their final destination on the same day (Fig. 32C). Migration continues on day E16 and is completed by day E17, when the massive facial nerve assumes its typical looped course as seen in the adult.

Comments. These observations indicate that on day E14, one day after the production of the neurons of the facial nucleus (96% originate by day E13; Altman and Bayer 1981b: Fig. 2), efferent fibers reach and move past the boundary cap of the facial nerve. This is comparable to what we observed in the trigeminal system, where efferents reach the trigeminal boundary cap on day E14 (Figs. 15 and 16). But in almost all other respects the chronology and course of events in the two systems are different: (a) the growth of facial efferents precedes perikaryal migration and the growth of trigeminal efferents follows (or is coupled with) cell migration; (b) facial efferents begin to grow one day after cell production and the growth of trigeminal efferents occurs 2 days thereafter (the production of trigeminal motor neurons is completed by day E12; Fig. 22); (c) facial cell migration begins on day E15, whereas the trigeminal

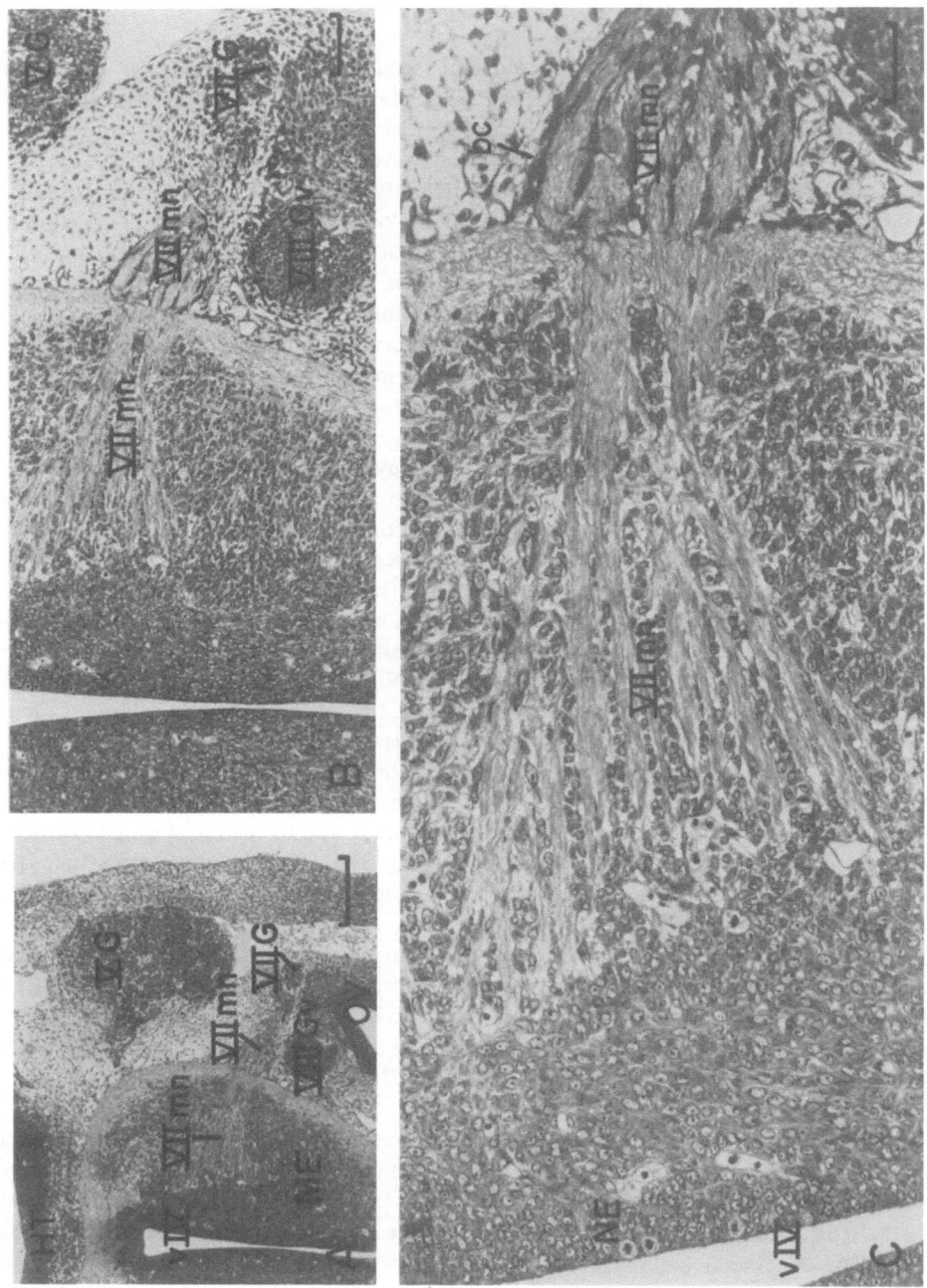

Fig. 28. *A* Horizontal section through the region of the medulla with fascicles of the facial efferents from a day E14 rat. *B* At higher magnification the fascicles are seen coursing in the direction of the facial ganglion. *C* Detail of the fascicles in the medulla and their aggregation into rootlets peripherally, where the fibers are surrounded by cells (presumably derivates of the facial boundary cap). Methacrylate. *Scales: A*, 300 µm; *B*, 100 µm; *C* 50 µm

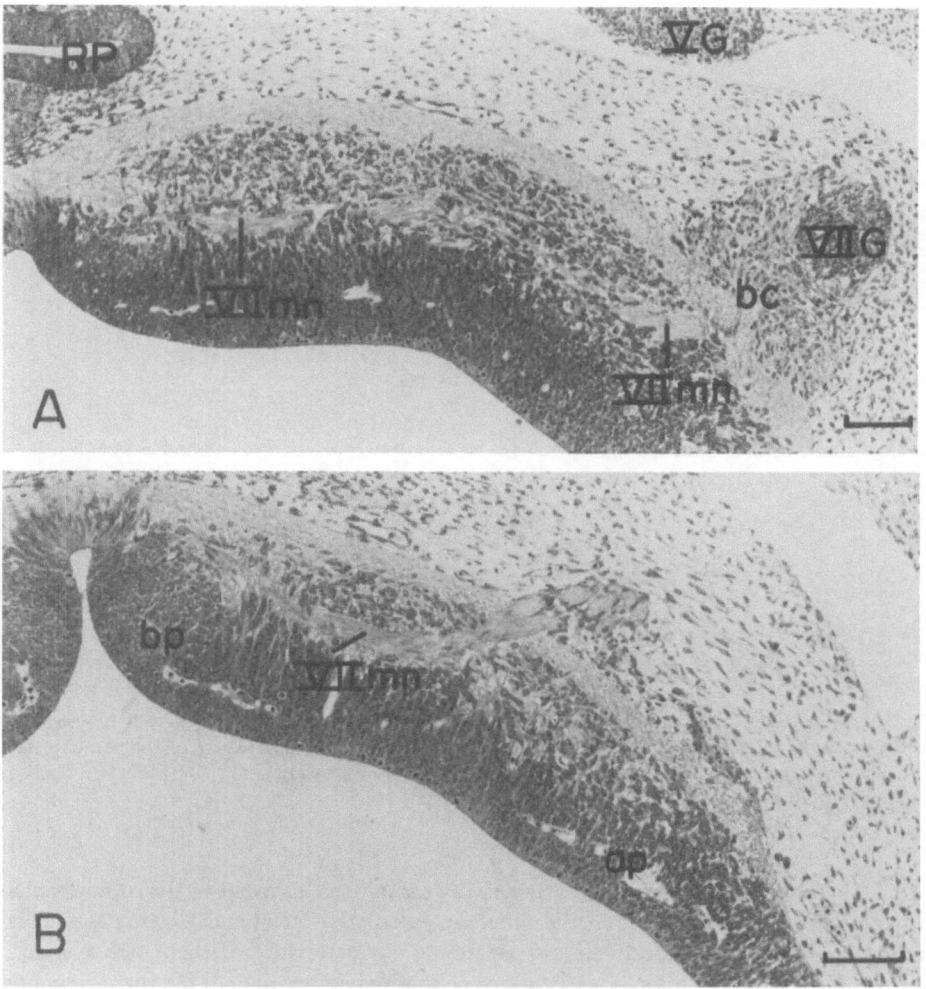

Fig. 29. *A* Coronal section from a day E14 rat showing the course of the fibers of the facial motor nerve in relation to the boundary cap of the facial ganglion. *B* In this section the entire course of the facial nerve is seen from the region of the medial neuroepithelium (considered the basal plate) in a lateral direction. Methacrylate. *Scale:* 100 μm

motor neurons begin to migrate on day E13; and finally, (d) the perikarya of the trigeminal motor neurons migrate in the direction of the trigeminal boundary cap, whereas the facial perikarya take an initial course directed at a right angle to their fibers and the facial boundary cap.

The fact that the facial efferents, like the trigeminal efferents, grow in the direction of the boundary cap, supports the hypothesis that this peripheral structure exerts a guiding influence on the central growth of motor fibers. Obviously, the boundary cap does not exert a similar influence on cell migration, since the motor neurons of the facial nerve move in a different direction. We shall discuss later what guiding influences

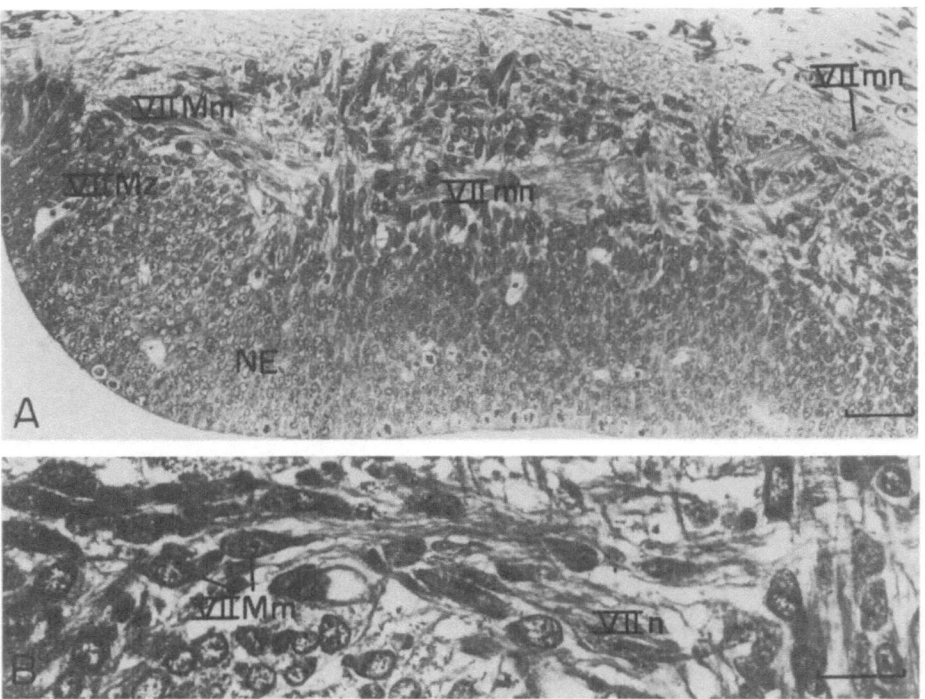

Fig. 30. *A* Coronal section from a day E14 rat. The flask-shaped differentiating cells that are the source of the facial efferents are identifiable medially in the superficial aspect of the neuroepithelium. *B* The flask-shaped cells at higher magnification with outgrowing axons. Methacrylate. *Scales: A*, 50 µm; *B*, 20 µm

may be involved in cell migration, after having examined this event in the other cranial nerve nuclei. The differences in the development of the facial and trigeminal ganglia are not surprising if we recall the organizational characteristics of these two systems. First, although both the facial and the trigeminal nerves are mixed, the major component of the trigeminal nerve is afferent and the major component of the facial nerve is efferent. Second, and more importantly, there must be an intimate relationship between trigeminal afferents and efferents but there is little, if any, between facial afferents and efferents. It may be assumed that the trigeminal motor neurons controlling muscles of the mandible receive input from afferents of the mandible. But as we noted above, it is not facial afferents but rather trigeminal afferents that are intimately involved in facial reflexes mediated by the facial motor nucleus.

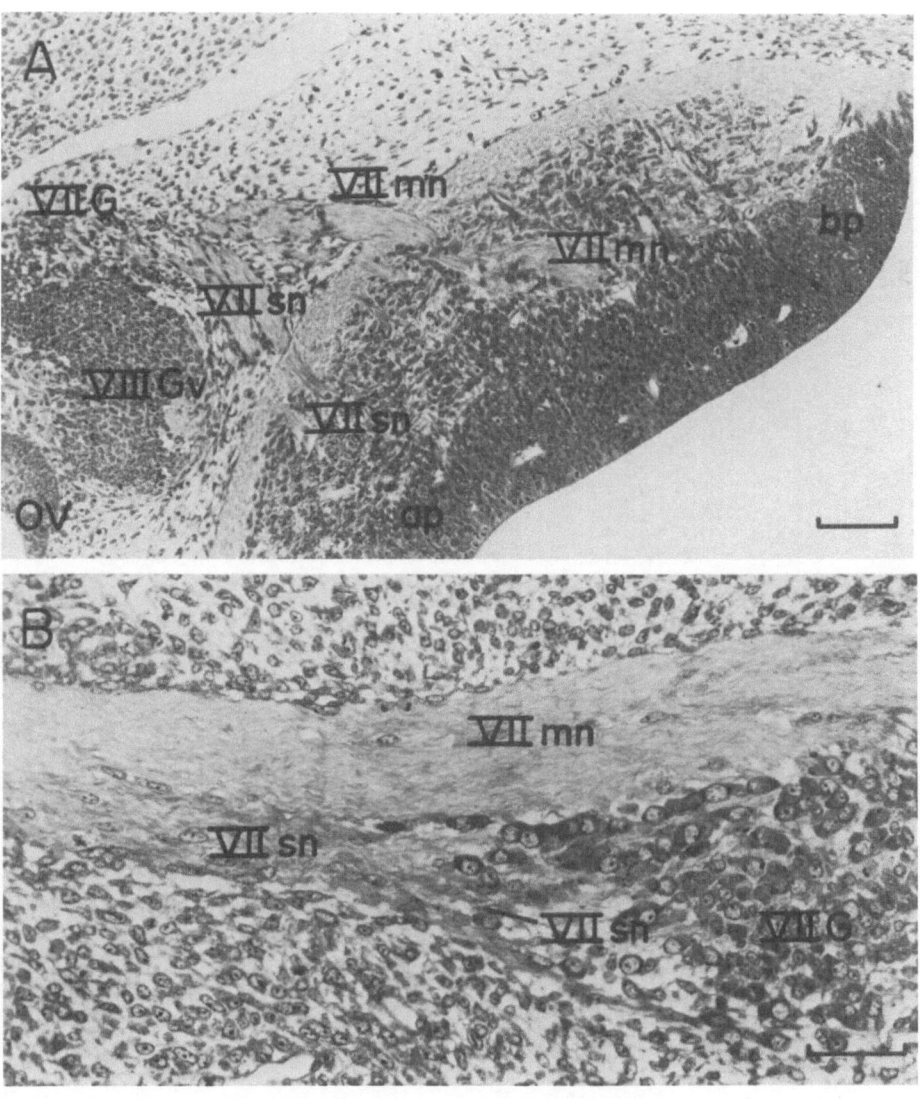

Fig. 31. *A* The sensory branch of the facial nerve (nervus intermedius) is seen entering the medulla in this day E14 rat. *B* In this section the facial motor nerve has grown past the facial ganglion where it is apparently joined by the distal facial afferents. Methacrylate. *Scales: A*, 100 μm; *B* 50 μm

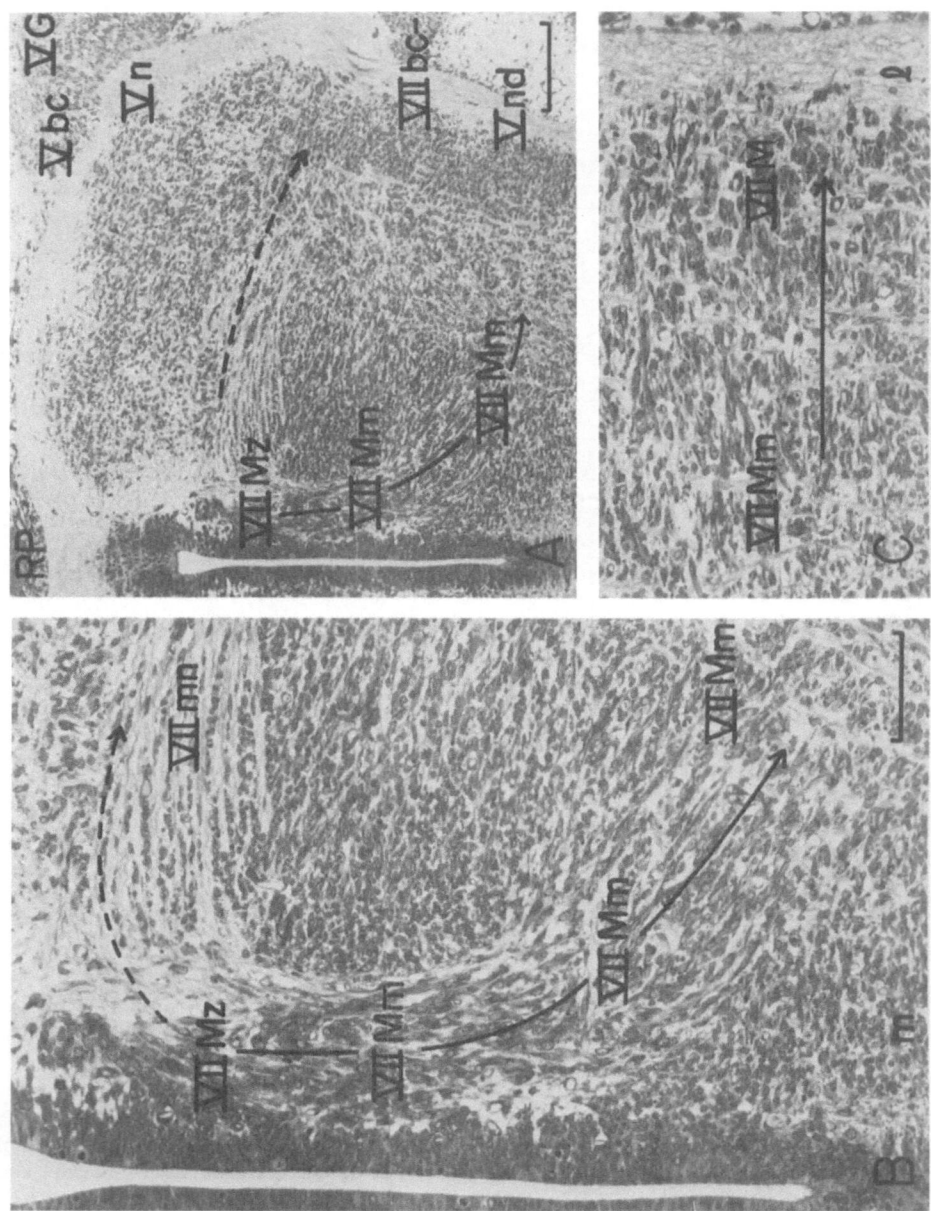

Fig. 32. *A* Horizontal section of the upper medulla from a day E15 rat. At this level the fascicles of efferents are coursing in the direction of the facial boundary cap *(broken arrow)*. Also seen are the facial motor neurons in apparent migration *(solid arrows)* from their medial site of production caudally and then laterally. *B* Detail of *A*. *C* In an adjacent section some facial motor neurons are seen ventrolaterally at the site of their final settling. Methacrylate. *Scales: A*, 200 μm; *B, C*, 100 μm

5 Development of the Vestibular Ganglion in Relation to the Vestibular Nuclei

5.1 Time of Origin of Vestibular Ganglion Cells

Background. The vestibular ganglion (the ganglion of Scarpa) contains the primary sensory neurons that convey information from the receptors of the labyrinth to the vestibular nuclei of the medulla. The distal (or peripheral) fibers of the pseudounipolar ganglion cells innervate the hair cells of the maculae of the utriculus and sacculus, and of the ampullary cristae of the semicircular ducts. The proximal (or central) fibers form the vestibular branch of the eighth nerve that enters the medulla. These fibers reach the lateral vestibular nucleus (the nucleus of Deiters) and divide into ascending and descending branches. The ascending fibers proceed to the superior vestibular nucleus, the descending fibers to the inferior (spinal or descending) and the medial vestibular nuclei (Truex and Carpenter 1969; Brodal 1981). Vestibular fibers are distributed to restricted parts of these nuclei and there is preferential projection from different components of the labyrinth to the different nuclei (Stein and Carpenter 1967; Gacek 1969).

To our knowledge there is no published information presently available on the time of origin of vestibular ganglion cells in any species. Relevant, perhaps, is the debated issue of the order of production of vestibular ganglion cells and spiral ganglion cells. On the basis of cytological observations in the pig, Campenhout (1935) claimed that the spiral ganglion cells form before the vestibular ganglion cells; in contrast, Batten (1958), who studied sheep, maintained that the development of the spiral ganglion cells lags behind the vestibular ganglion cells.

Results. The proportion of labeled to unlabeled vestibular ganglion cells was determined in thymidine radiograms sectioned in the sagittal plane (Fig. 1). In rats injected either on days E11+12 or days E12+13, all the ganglion cells were labeled. In the latter group, label concentration was intermediate in most ganglion cells but a few cells were heavily labeled. In the E13+14 group, only 71% of the cells were labeled, indicating that 29% of the cells ceased to multiply during the period between the mornings of day E12 and E13. Most of the unlabeled cells were large neurons. In the E14+15 group only 3% of the cells were labeled, mostly, though not exclusively (Fig. 33A), smaller neurons. In this injection group (Fig. 33B), and in all the groups up to days E17+18, the numerous satellite cells of the vestibular ganglion were consistently labeled. In contrast to the vestibular ganglion cells, a very high proportion of the spiral ganglion cells (Fig. 34) were labeled in the E14+15 group. We conclude, first, that neurons of the vestibular ganglion arise between days E12 and E14, with 97% of them being generated on days E12+13 (Fig. 35); and second, that the ganglion cells of the vestibular system are produced ahead of the ganglion cells of the cochlea.

Comments. Our evidence that the neurons of the vestibular ganglion are generated before the spiral ganglion cells supports Batten's (1958) inferences as against Campenhout's (1935). Our results also revealed that the ganglion cells of the labyrinth are produced later (peak on day E13) than the facial ganglion cells (peak on day E12; Fig. 21). While the peak of cell production of vestibular ganglion cells was similar to that of the trigeminal ganglion, the latter were produced over a longer time span (days E11–E15; Fig. 5).

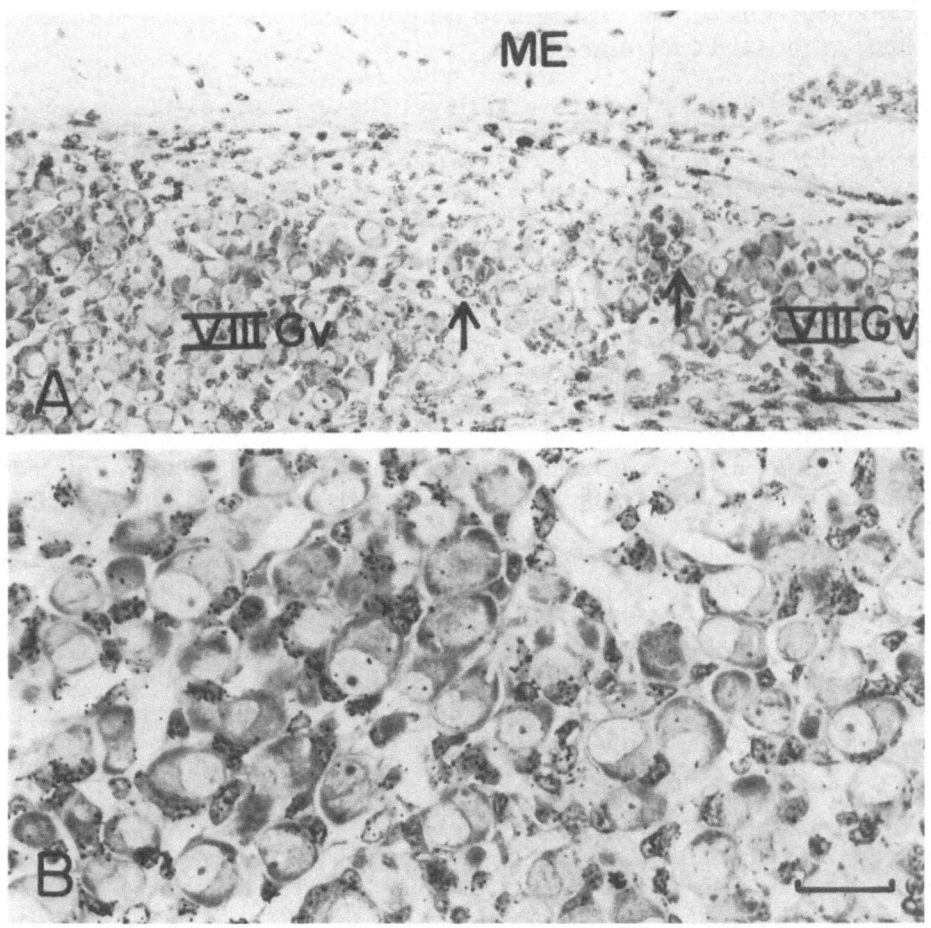

Fig. 33. *A* Pattern of cell labeling in the vestibular ganglion of a rat injected on days E14+15. *Arrows* point to the few labeled cells. *B* Detail at higher magnification to show the labeling of satellite cells. Methacrylate. *Scales: A*, 50 μm; *B*, 20 μm

5.2 Embryonic Development of the Vestibular Ganglion

Background. The embryonic development of the vestibular ganglion (or more generally of the "auditory" or "otic" ganglion) has been the subject of controversies: first, whether the otic ganglion and the facial ganglion are derived from a shared primordium; and, second, whether the otic ganglion is of neural crest derivation or placodal origin. On the one side was Adelmann, who wrote: "The VIII ganglion is a derivative of the common acoustico-facial ganglionic mass owing its origin entirely to neural crest proliferation" (Adelmann 1925, p. 118). An intermediate position was adopted by Halley (1955), who accepted the idea of a common acousticofacial primordium but maintained that the otic ganglion arises at least in part from the otic vesicle. At the other end was Batten (1958), who concluded that the otic and facial ganglia originate from independent cell masses and not from a common primordium, and that the otic

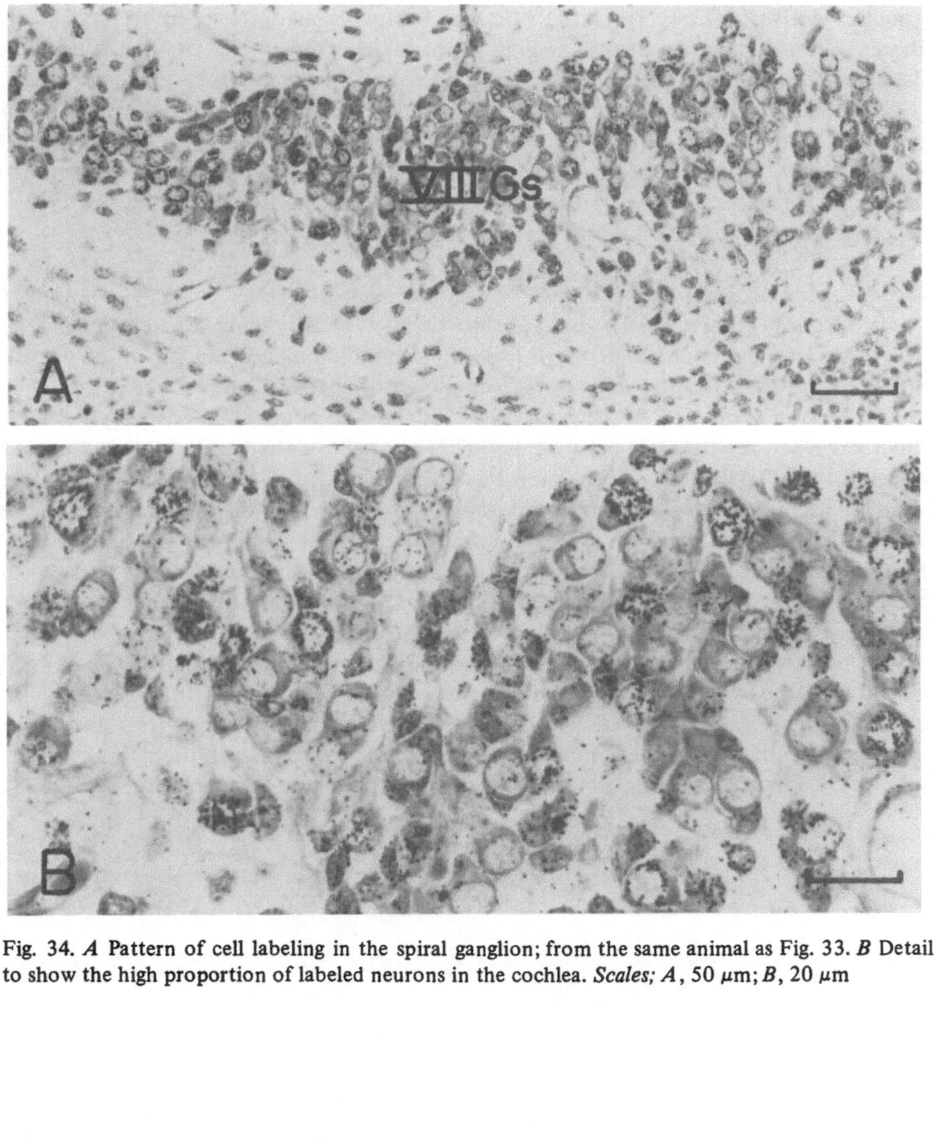

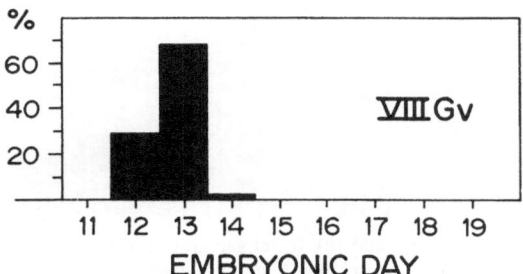

Fig. 34. *A* Pattern of cell labeling in the spiral ganglion; from the same animal as Fig. 33. *B* Detail to show the high proportion of labeled neurons in the cochlea. *Scales; A*, 50 μm; *B*, 20 μm

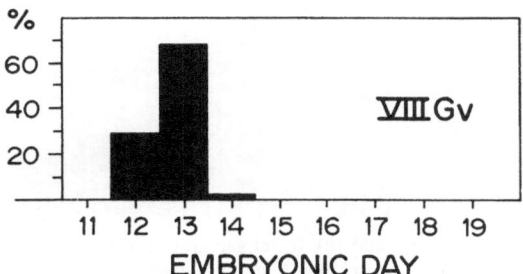

Fig. 35. Time of origin of neurons in the vestibular ganglion

ganglion derives entirely from the otic vesicle. In his material in embryonic sheep, Batten observed streams of cells issuing from the otic vesicle and assumed that these detached cells were the specific source of the neurons of the auditory ganglia.

In our preceding study of the development of the facial ganglion, we found support for Batten's view of the separate origins of the facial and otic ganglia. In the present study, in line with our observations of the sequential production of vestibular and spiral ganglion cells, we present observations that the two ganglia are derived in succession from the epithelium of the otic vesicle.

Results. Invagination of the otic vesicle begins on day E11 (Figs. 6, 23), at the time of onset of somite formation. The base of the otic pit typically approximated the medulla at the level of rhombomere 4 (Fig. 23A). The rounding-up and detachment of the otic vesicle from the skin primordium was completed in all the examined rats by day E12 (Fig. 24). The formation of the primordium of the otic ganglion (specifically the vestibular ganglion; see below) began on day E12. The cells of the vestibular ganglion were apparently detaching themselves from the rostral and rostrolateral aspect of the dorsal wall of the otic vesicle, at intermittent points (Fig. 36) where the basement membrane seemed to be absent. In this way the primordium of the vestibular ganglion became inserted between the otic vesicle caudally and the facial ganglion rostrally (Fig. 24A). In well-preserved material the differentiating cells of the facial ganglion could be distinguished from the primitive cells of the vestibular ganglion (Fig. 36).

The detachment of the vestibular ganglion cells from the wall of the otic vesicle continued on day E13 (Fig. 37). Mitotic cells were frequent in the formative ganglion (Fig. 37A), suggesting that the neurons generated on this day were at least partly produced regionally. In day E13 embryos it was possible to distinguish the laterally situated vestibular ganglion from the medial boundary cap close to the medulla (Fig. 38).

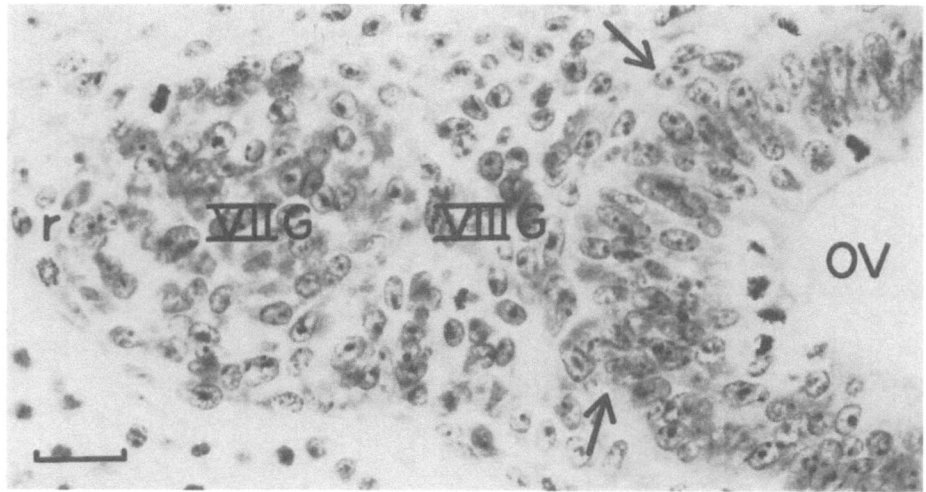

Fig. 36. The otic vesicle, rudiments of the otic ganglion, and the facial ganglion in a day E12 rat. *Arrows* point to regions in the rostral otic vesicle where cells may become detached to form the otic ganglion. The facial ganglion is recognizable as a spherical, separate structure. Paraffin. *Scale:* 20 μm

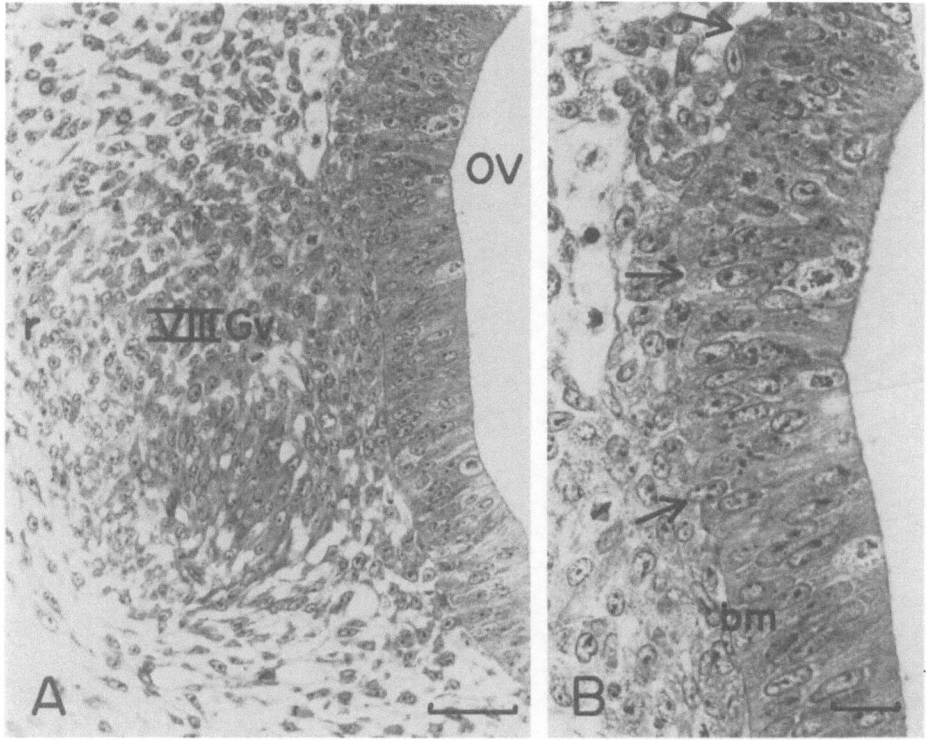

Fig. 37. *A* The rudiment of the vestibular ganglion in relation to the otic vesicle in a day E13 rat. *B* Detail to show absence of the basement membrane at intermittent points *(arrows)* where cells of the otic vesicle migrate into the vestibular ganglion. Methacrylate. *Scale: A*, 50 μm; *B*, 20 μm

The vestibular boundary cap was traversed by the earliest contingent of proximal vestibular afferents, which penetrated the medulla at several points (Fig. 39).

In day E14 rats, the distal fibers of the vestibular nerve could be followed from the ganglion around the medial wall of the otic vesicle in a caudal direction (Fig. 40). There were no indications that, at this age, the fibers penetrated the tissue of the otic vesicle. While vestibular neuron production came to a virtual end on day E13 (Fig. 35), a new stream of cells issued from the anteroventral aspect of the otic vesicle on day E14 (Figs. 41–42). On the basis of our thymidine-radiographic studies and embryonic observations made in day E15 rats (Fig. 43), we identify this second stream of cells originating in the otic vesicle as the precursors of spiral ganglion cells.

The spots where the basement membrane of the otic vesicle were absent, and where cells were parting from its wall, became restricted by day E15 (Fig. 43). But many primitive, proliferating cells were still present in the vicinity of the otic vesicle. The distal cells were beginning to differentiate and formed a coil rostrally. The difference in the degree of maturation of vestibular ganglion cells and the latest-generated spiral ganglion cells was appreciable. The differentiating vestibular ganglion cells typically had one or two densely staining apical cytoplasmic cones, presumably reflecting intense metabolic activity associated with the production of fibers. Mitotic cells were not seen at this age in the ganglion, suggesting that the precursors of satellite cells are not produced locally.

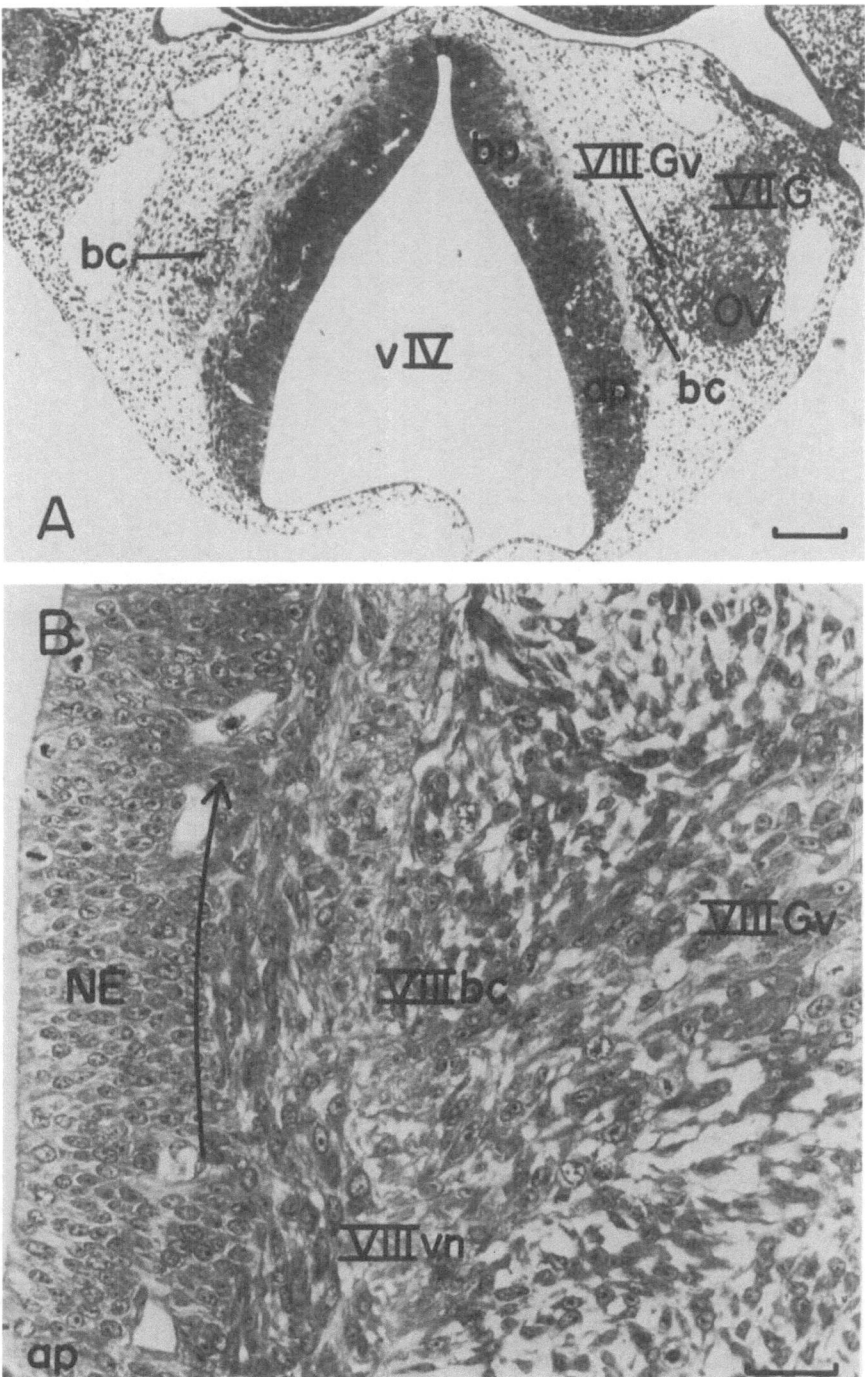

Fig. 38. *A* Horizontal section from a day E13 rat showing the boundary cap of the vestibular ganglion in the vicinity of the alar plate. *B* Detail of right side in *A* showing fibers traversing the boundary cap. *Vertical arrow* is drawn parallel to the apparently migrating cells from the region of the alar plate. Methacrylate. *Scales: A*, 150 μm; *B*, 100 μm

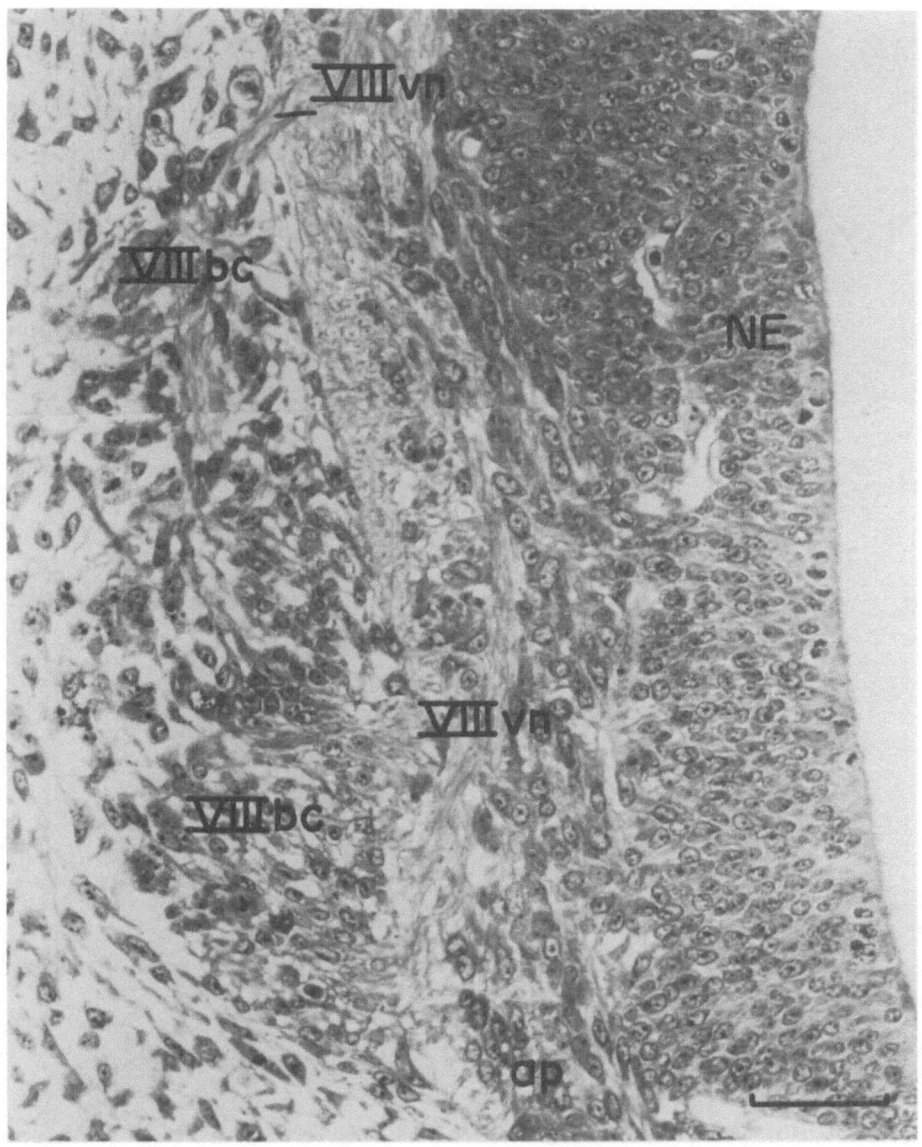

Fig. 39. Detail of the left side in Fig. 38 A, showing vestibular fibers penetrating the medulla on day E13. *Scale:* 50 μm

Comments. The idea of the separate origin of the otic and facial ganglia, originally proposed by Streeter (1906) and later championed by Batten (1958), gains support from three lines of evidence in the present study. First, our thymidine-radiographic results indicate that the facial ganglion cells are generated earlier (peak on day E12) than the vestibular ganglion cells (peak on day E13), and at a time when the detachment of cells from the otic vesicle has just begun. Second, our embryonic observations indicated that the facial rudiment is related to a placode in the hyoid arch region,

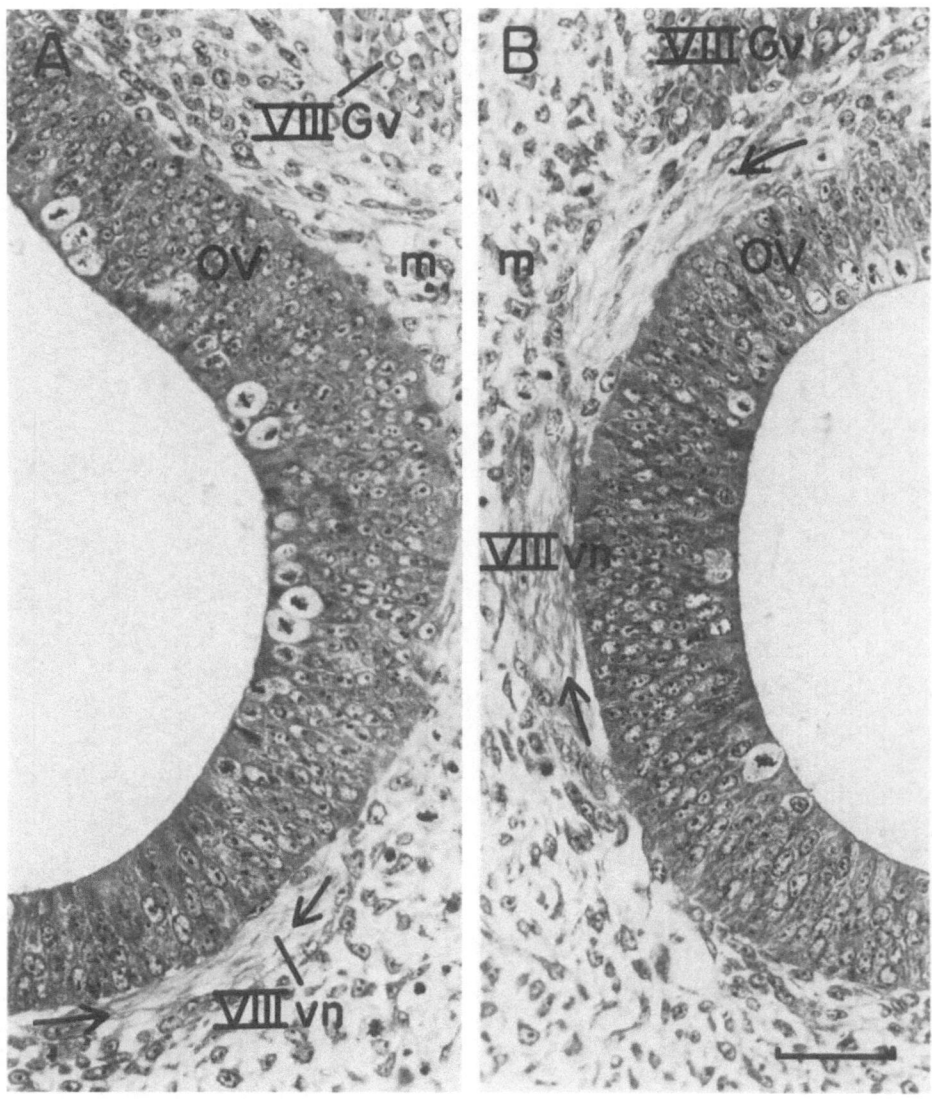

Fig. 40. Distal vestibular fibers coursing along the medial wall of the otic vesicle *(arrows* in *A* and B). The fibers can be traced to the caudal aspect of the otic ganglion *(A)*. Both from a single section in a day E14 rat. Methacrylate. *Scale:* 50 μm

whereas the vestibular and spiral ganglion are related to the otic vesicle. Third, the facial ganglion was a recognizable discrete structure with differentiating cells as early as day E12, that is, at a time when the detachment of the first wave of cells from the otic vesicle (presumably the precursors of vestibular ganglion cells) has only just started and when the detachment of the second wave (the presumed precursors of spiral ganglion cells) has not yet begun. Insofar as the otic vesicle is a placodal structure (it forms through the invagination of a superficial ectodermal thickening), the present

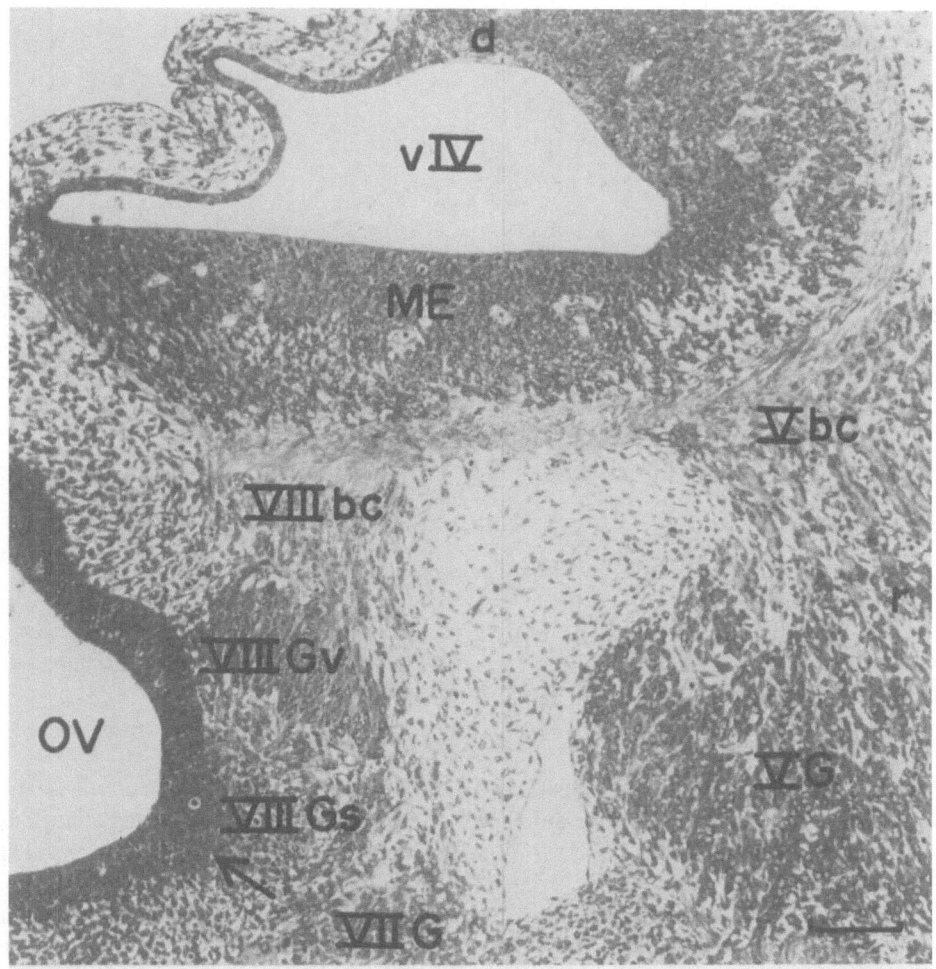

Fig. 41. Sagittal section from a day E14 rat showing the position of cells, identified as precursors of the spiral ganglion cells, issuing from the otic vesicle *(arrow)*. Methacrylate. *Scale:* 100 μm

evidence also supports the idea of the placodal origin of cranial nerve ganglia. However, as in the case of the trigeminal ganglion, it is possible that the regional boundary cap is of neural crest derivation.

5.3 Embryonic Development of the Vestibular Nuclei

Background. According to our radiographic datings (Altman and Bayer 1980c: Fig. 1), the neurons of the vestibular nuclei are generated sequentially between days E11 and E15; the lateral (Deiter's) nucleus with a peak on day E12; the superior nucleus with a peak on day E13; the inferior nucleus with a peak on day E13 but a high proportion of cells produced on day E14; and the medial nucleus with a peak on day E14. These

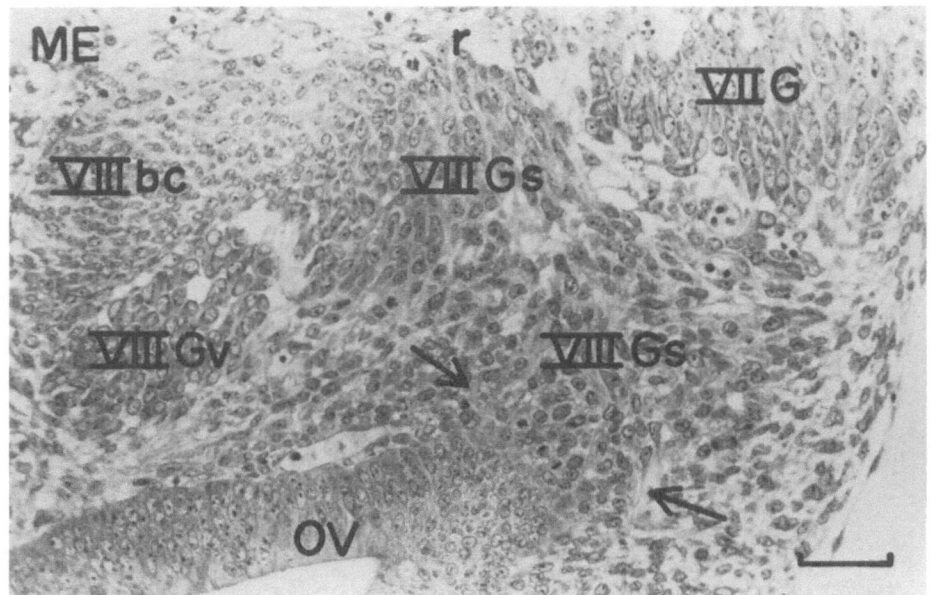

Fig. 42. Horizontal section from a day E14 rat. *Arrows* point to the site where precursors of the spiral ganglion leave the wall of the otic vesicle. Cells appear more differentiated in the distal portion of the spiral ganglion rostrally. Methacrylate. *Scale:* 50 µm

results suggested two lateral-to-medial cytogenetic gradients: rostrally, from the lateral nucleus to the superior nucleus; and caudally, from the inferior to the medial nucleus (Altman and Bayer 1980c: Fig. 18). We postulated that the vestibular nuclei represent a single cytogenetic zone, called "zone VE," with the implication that it has two components, an earlier-forming rostral division and a later-forming caudal division (Altman and Bayer 1980c: Fig. 19). In the present embryological study we sought to identify the neuroepithelial zone from which the vestibular nuclei may be derived, and correlate its development with concurrent peripheral events.

Results. We located in day E13 rats (Fig. 38) a stream of spindle-shaped cells between the posterolateral flank of the rhombencephalic neuroepithelium (the alar plate) and the vestibular boundary cap. The site of this migratory stream parallels and is contiguous with the first contingent of afferents of the vestibular ganglion (Fig. 39). Since over 65% of the neurons of the lateral vestibular nucleus have left the neuroepithelium by the morning of day E13, we assume that these cells are the migrating Deiters neurons that have left the alar plate on their route toward the vestibular boundary cap. The migratory stream continues on day E14, possibly consisting of the remaning neurons of the lateral nucleus and of the neurons of the superior nucleus that are generated with a peak on day E13.

In order to understand the subsequent developmental events, we have to consider a regional morphological change that becomes evident by day E15 (Fig. 44) and continues thereafter. This is the formation of the lateral recess of the fourth ventricle. The event is heralded by the enlargement of the fourth ventricle on day E14 (Fig. 44B),

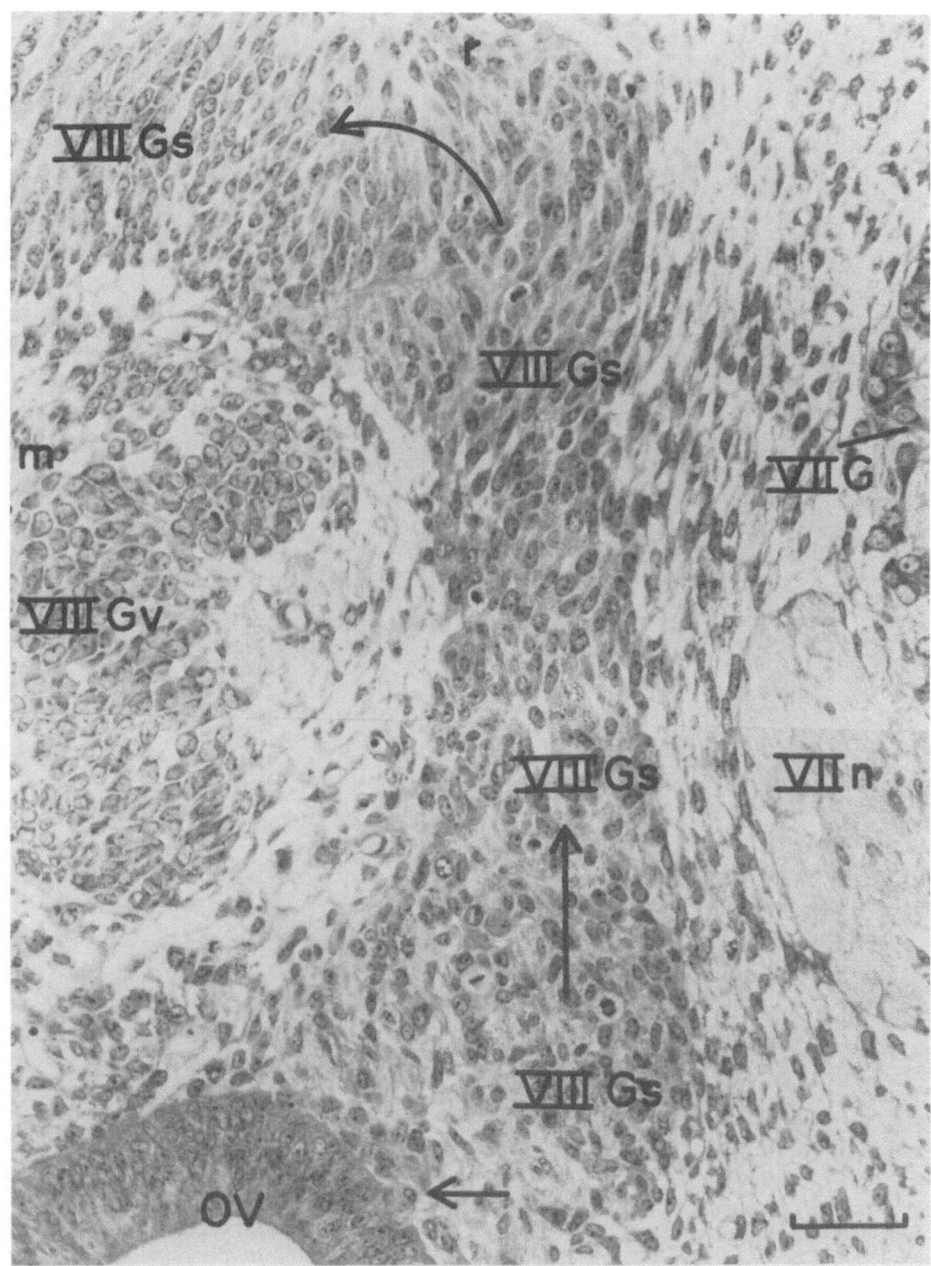

Fig. 43. Horizontal section from a day E15 rat. The spiral ganglion has considerably lengthened and coils rostrally in a medial direction. Few cells are leaving the otic vesicle *(small horizontal arrow)*, but mitotic activity is still high in the formative spiral ganglion in the region of the otic vesicle. Methacrylate. *Scale:* 50 μm

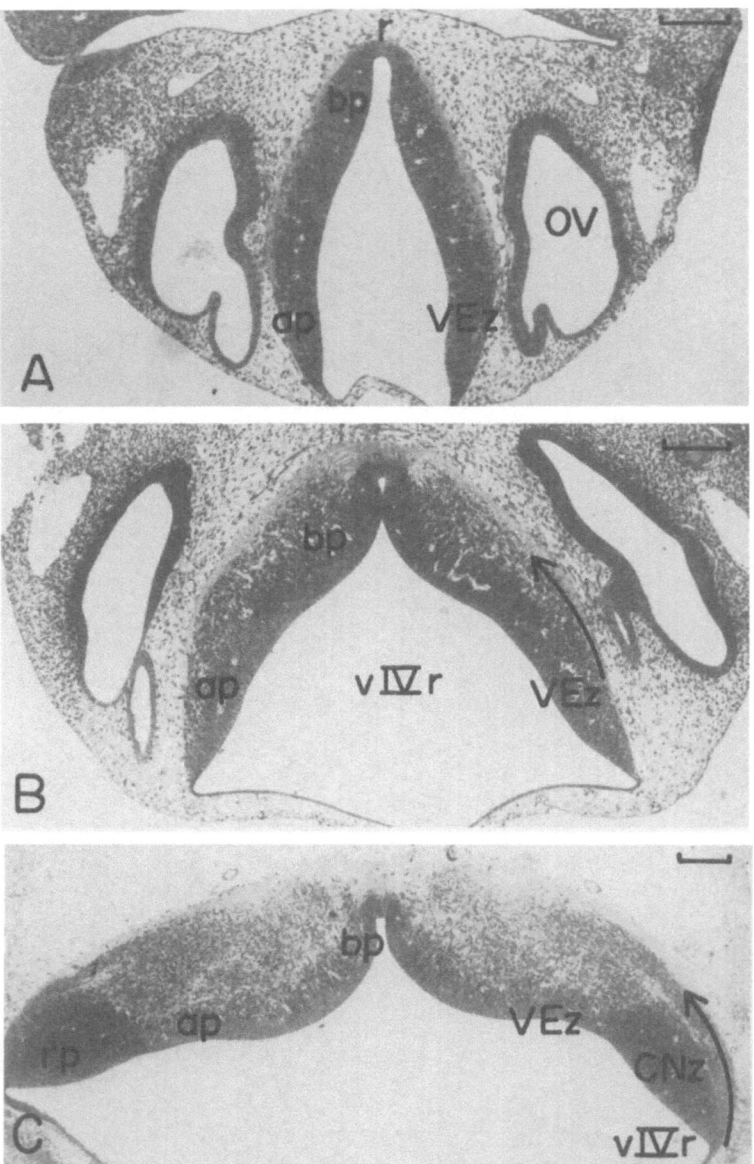

Fig. 44 A–C. Formation of the recess of the fourth ventricle and the recess plate of the neuro-epithelium of the cochlear nuclear complex. *A* The vestibular neuroepithelium occupies a lateral position in a day E13 rat. *B* The first indications of the formation of the lateral recess of the fourth ventricle on day E14. *Arrow* outlines cell migration (vestibular) from the alar plate. *C* The vestibular neuroepithelium is displaced medially by the enlarged cochlear neuroepithelium on day E15. *Arrow* outlines cell migration (cochlear) from recess plate. Methacrylate. *Scales: A–C,* 200 μm

brought about partly by an outward movement of the rhombencephalic walls. By day E15 (Fig. 44C) the recess of the fourth ventricle has started to form as a result of the addition of a large neuroepithelial mass laterally, the recess plate. The recess plate is the primordium of the cochlear nuclear complex, all components of which (the antero-ventral, posteroventral, and dorsal nuclei) acquire most of their neurons on day E15 (Altman and Bayer 1980c: Fig. 7). As a result of this change the previously laterally situated vestibular neuroepithelium (the alar plate) now assumes an intermediate position between the recess plate and the basal plate.

We do not assume that the cochlear neuroepithelium begins to form on day E15, since the pyramidal cells of the dorsal cochlear nucleus are generated as early as days E12–E14 (Altman and Bayer 1980c). Rather, the cochlear neuroepithelium becomes conspicuous on day E15 because the spurt in its proliferative activity on this day is coupled with the sudden shrinkage of the adjacent portion of the vestibular neuro-epithelium (Fig. 45). We located two clusters of differentiating neurons in the vicinity of this regressive germinal matrix (Fig. 45B). We assume that the large-celled lateral group is the earlier lateral vestibular nucleus and that the adjacent cluster of smaller cells is the later-forming (and ultimately smaller-celled) superior vestibular nucleus. On the basis of these considerations, and in line with our earlier dating studies, we designate the lateral portion of the alar plate at this level of the neuraxis as subdivision 1 of the vestibular neuroepithelial zone (VE, z1). The medial aspect of this neuro-epithelial zone, designated as subdivision 2 (VE, z2), began to regress on day E16.

Comments. Our embryonic investigations in this phase of our inquiry had two aims: (a) to relate the chronology of events in the vestibular nuclei centrally to events in the vestibular ganglion peripherally; and (b) to identify the neuroepithelial sources (referred to as neuroepithelial zones) of the vestibular nuclei neurons whose time of origin was determined in an earlier study (Altman and Bayer 1980c).

Coinciding with the penetration of vestibular afferents into the medulla, we observed in day E13 rats a stream of migratory cells coursing in the same direction from the region of the alar plate. In view of our evidence that most Deiters cells are produced on day E12, and that most neurons of the other vestibular nuclei are generated later with a peak on day E13 or E14 (Altman and Bayer 1980c; Fig. 1), we identified this stream of cells as the young Deiters neurons. The vestibular primary afferents are said to distribute to the lateral vestibular nucleus before bifurcating into ascending and descending branches and proceeding to the other vestibular nuclei; preparatory steps for this initial contact may then begin on day E13. The convergent growth of peripheral vestibular afferents and migration of central sensory neurons suggests the possibility of some reciprocal morphogenetic interaction between vestibular afferents and their recipient nerve cells.

In the light of our datings of the time of origin of neurons of the four vestibular nuclei (Altman and Bayer 1980c), we were able to identify with some assurance two components of the vestibular neuroepithelium, and distinguish these from the cochlear neuroepithelium. The key to this identification was the evidence that the bulk of the neurons of the lateral and superior vestibular nuclei are produced on days E12–E13, the bulk of the neurons of the inferior and medial vestibular nuclei on days E13–E14 (Altman and Bayer 1980c: Fig. 1), whereas peak formation time of neurons of the anteroventral, posteroventral, and dorsal cochlear nuclei is on day E15 (Altman and Bayer 1980c: Fig. 7). In day E15 rats we could distinguish three neuroepithelial

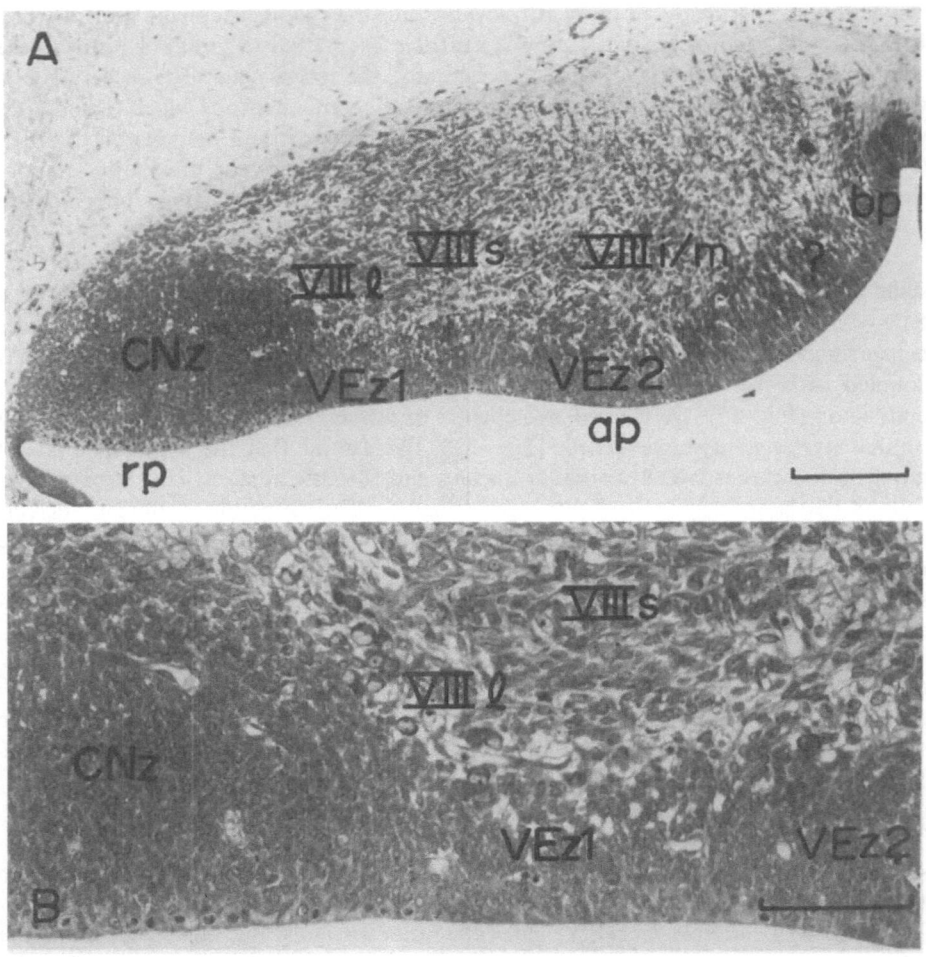

Fig. 45. *A* Presumed subdivisions of the neuroepithelium in a day E15 rat. *B* Detail of the region of the receding vestibular neuroepithelium, zone 1, with the maturing Deiters neurons and younger neurons of the superior vestibular nucleus. Methacrylate. *Scales: A*, 200 µm; *B* 100 µm

regions in the medulla in the vicinity of the inner ear. First, there was a far-lateral, extremely active zone in the region of the recess of the fourth ventricle; we designated this as the cochlear neuroepithelium. Second, adjacent to the cochlear neuroepithelial zone was a receding germinal zone, in the vicinity of which we noted the large Deiters neurons and another aggregate of smaller cells. We designated this germinal strip as the original site of origin of the large neurons of the lateral and superior vestibular nuclei. Finally, medial to the latter region was a somewhat more active germinal zone with less distinct aggregates of cells beneath it; it was presumed that this region is the germinal source of neurons of the inferior and medial vestibular nuclei.

6 Development of the Glossopharyngeal and Vagal Ganglia in Relation to the Solitary and Ambiguus Nuclei

6.1 Time of Origin of Glossopharyngeal and Vagal Ganglion Cells

Background. Because of the close structural and functional association of the glosso-pharyngeal and vagal ganglia and their nerves, the two are usually considered together. Both are composed of two parts, a superior (or root) ganglion and an inferior (or trunk) ganglion. The inferior ganglion of the glossopharyngeal nerve (IXGi) is commonly referred to as the petrosal (petrous); there is no special term for the superior glossopharyngeal ganglion (IXGs). The superior ganglion of the vagus (XGs) is known also as the jugular, and its inferior ganglion (XGi) as the nodose. The afferents of IXGs and XGs are classified as somatic (or cutaneous) and are believed to project mainly to the spinal trigeminal nuclei; the afferents of IXGi and XGi are classified as visceral and they project mainly to the solitary nucleus.

According to textbook descriptions (e.g., Truex and Carpenter 1969; Brodal 1981), the somatic afferents of IXGs and XGs innervate the cutaneous area in the back of the ear, jointly forming the auricular branch of the vagus nerve. These somatic fibers may be those that have been traced centrally to the spinal nucleus of the trigeminal nerve (Torvik 1956; Kerr 1962; Cottle 1964). The visceral afferents of IXGi innervate the mucous membranes of the posterior third of the tongue (including the taste buds), the tonsil, and the eustachian tube. The visceral afferents of XGi innervate the pharynx, larynx, trachea, esophagus, a few scattered taste buds in the region of the epiglottis, and the thoracic and abdominal viscera. The visceral afferents of the vagus terminate in the solitary nucleus caudally, those of the glossopharyngeal more rostrally (Torvik 1956; Kerr 1962; Cottle 1964).

The glossopharyngeal and vagal nerves also contain efferents. The efferents that reach the muscles of the larynx, pharynx, and upper part of the esophagus originate from the large motor neurons of the ambiguus nucleus (Szabo and Dussardier 1964; Lawn 1966). The efferents of the glossopharyngeal nerve may originate from a collection of cells anterior to the ambiguus nucleus proper, which we have referred to as the retrofacial nucleus (Altman and Bayer 1980a). The efferents of the ambiguus nucleus contribute mainly to the vagus nerve (Truex and Carpenter 1969; Brodal 1981). Another source of vagal efferents are the preganglionic fibers that originate in the spindle-shaped cells of the dorsal motor nucleus of the vagus. Physiological studies have implicated the solitary nucleus in taste perception (Blomquist and Antem 1965) and, together with the ambiguus nucleus and the dorsal nucleus of the vagus, in swallowing (Roman and Car 1967; Car and Roman 1970) and in cardioinhibitory functions (Gunn et al. 1979). There are, to our knowledge, no reports available about the time of origin of the glossopharyngeal and vagal ganglia.

Results. The locations of IXGi, IXGs, XGi, and XGs are illustrated in a sagittal thymidine radiogram in Fig. 1. At higher magnification (Fig. 46) it is evident that in this rat labeled on days E14+15, the inferior or trunk ganglia (IXGi and XGi) contain few labeled neurons, but many neurons are labeled in the superior or root ganglia (IXGs and XGs). The numerous labeled neurons in the superior ganglia ranged from small to

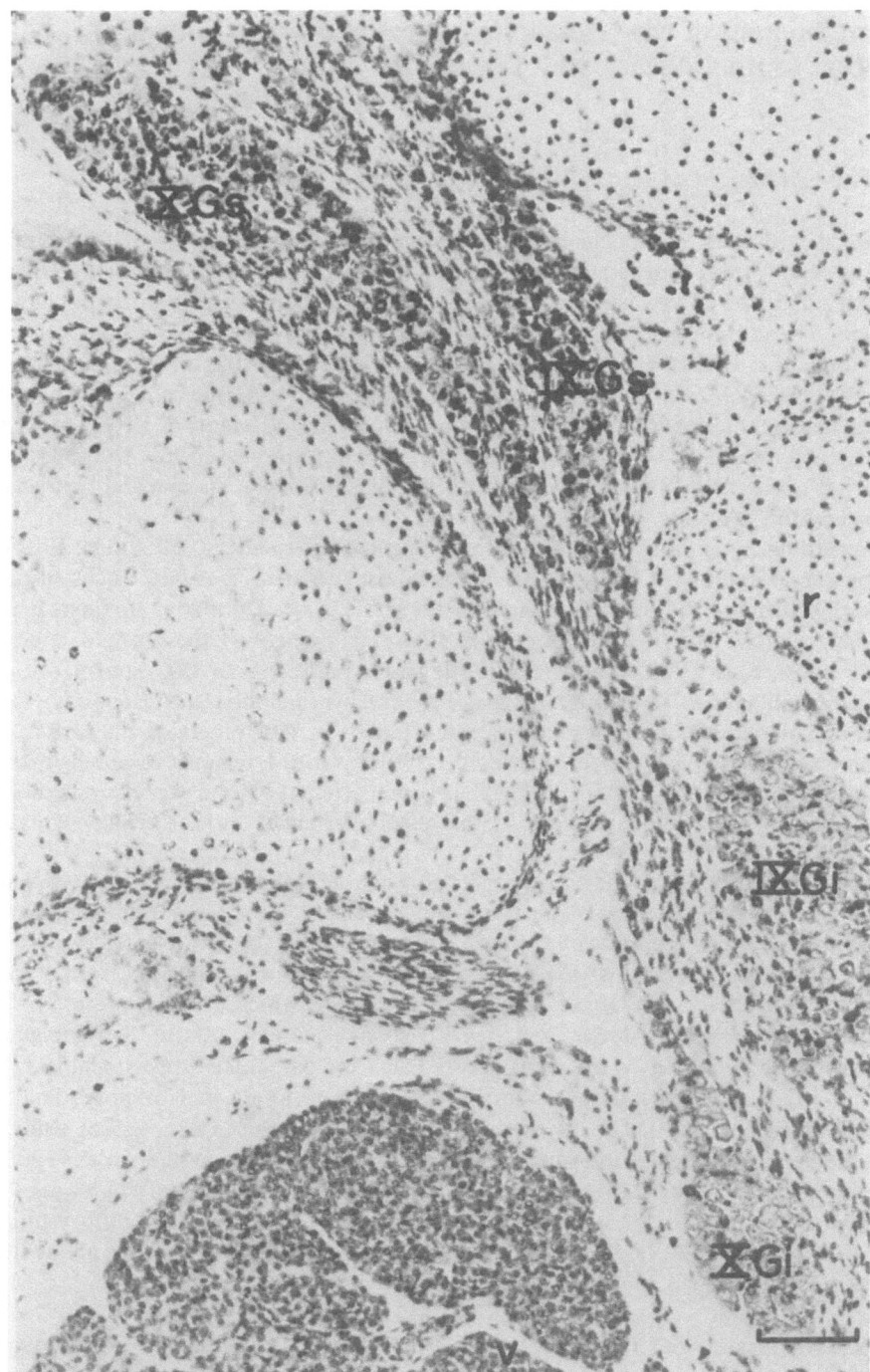

Fig. 46. Thymidine radiogram of the region of the glossopharyngeal and vagal ganglia in sagittal section from a rat injected on days E14+15 (compare with Fig. 1). Many neurons are labeled in the superior ganglia but only a few in the inferior ganglia (in the latter, most of the labeled elements are satellite cells). Paraffin. *Scale:* 100 μm

54

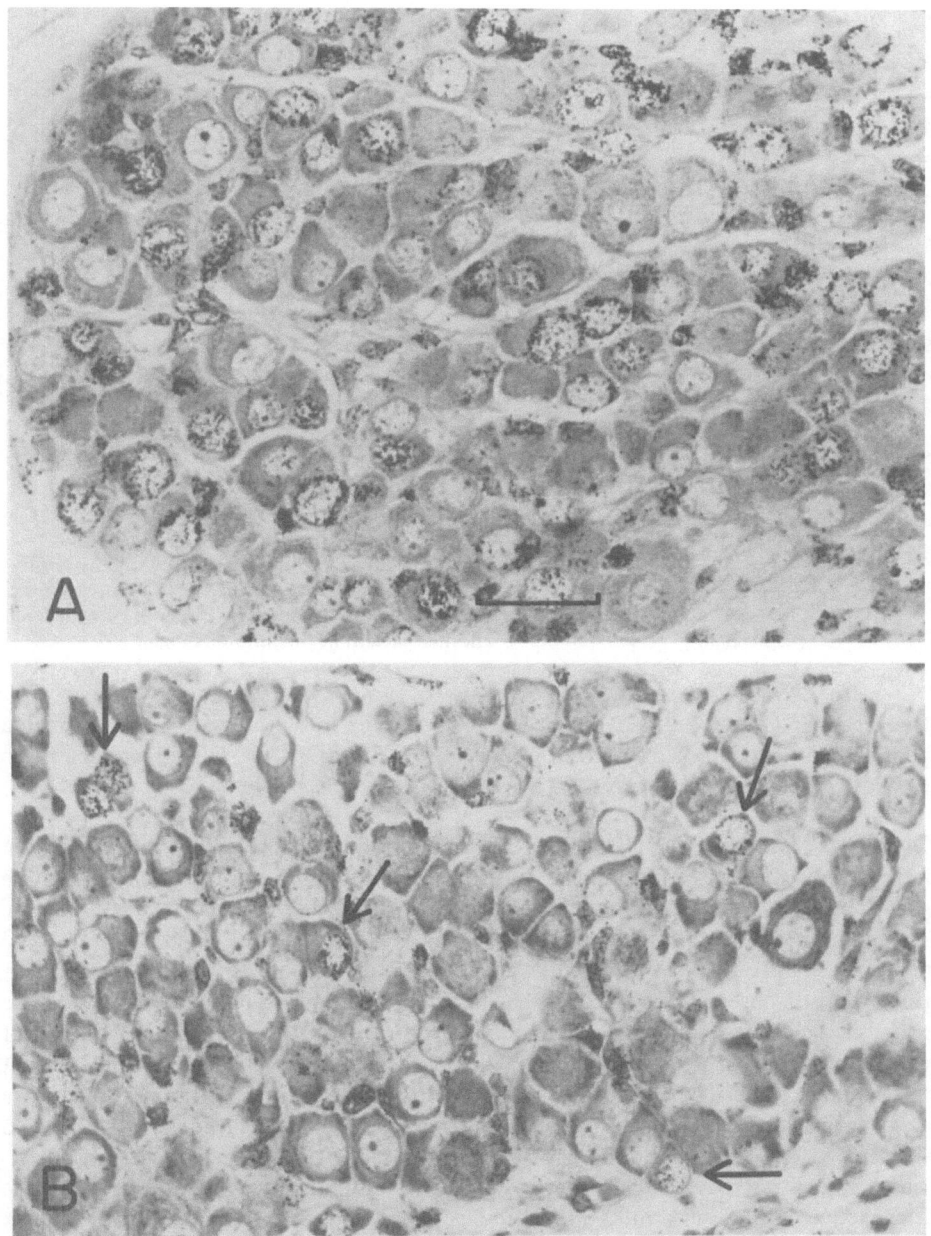

Fig. 47. *A* Illustration of the pattern of cell labeling in the superior ganglia and *B* in the inferior ganglia in a rat labeled on days E14+15. *Arrows* in *B* point to the labeled neurons. Methacrylate. *Scale:* 20 μm

large (Fig. 47A); the occasional labeled neurons in the inferior ganglia tended to be of the small type (Fig. 47B). We could not detect a spatial gradient of labeling within the individual ganglia. The quantitative results (Fig. 48) showed that the neurons of the

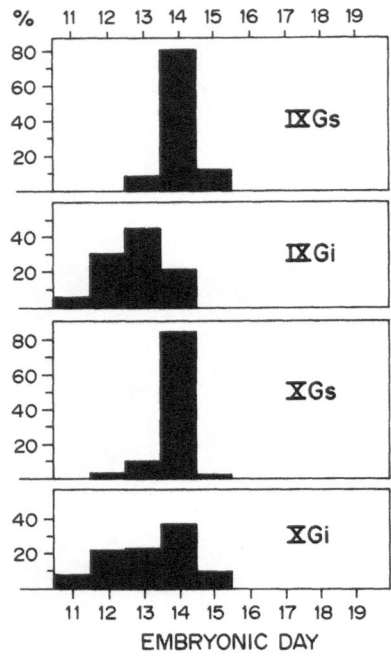

Fig. 48. Time of origin of neurons in the glosso-pharyngeal and vagal ganglia. Note one pattern of cell production in the two superior ganglia and another pattern in the inferior ganglia

trunk ganglia (IXGi and XGi) formed not only earlier but also over a more protracted period than the neurons of the root ganglia (IXGs and XGs). The latter showed a sharp peak of cell production on day E14 when 80% of the cells originate.

Comments. The pattern of ganglion cell production obtained in this study might be interpreted to indicate that there is a ventral-to-dorsal cytogenetic gradient within the glossopharyngeal ganglia (IXGi → IXGs) and also within the vagal ganglia (XGi → XGs). This would imply, as might be expected, that the two ganglia of the glossopharyngeal nerve represent one cytogenetic unit, and the two ganglia of the vagal nerve another. However, our observations indicate that both vagal and glossopharyngeal nerves share components of two different cytogenetic zones. The superior ganglia have a different pattern and time of neurogenesis from those of the inferior ganglia. This suggests that the superior ganglia come from a common germinal zone, while the inferior ganglia come from another. This interpretation is reconcilable with the evidence that the two superior ganglia are structurally and functionally related as sources of cutaneous affe-rents, in contrast to the two inferior ganglia, which are sources of visceral afferents. Irrespective of these interpretations, the results indicate that in IX–X nerve complex the ganglion cells of visceral afferents are produced before the ganglion cells of somatic afferents.

6.2 Embryonic Development of the Glossopharyngeal and Vagal Ganglia

Background. The study of the embryonic development of the glossopharyngeal and vagal ganglia has been a relatively neglected subject. According to Adelmann (1925), a continuous postotic neural crest is the primordium of the IX and X ganglia in the rat. The anlage was first recognized in eight-somite embryos; the separate IX and X ganglia could be distinguished in 24-somite embryos; and in 34-somite embryos the superior and inferior components of each ganglion could be recognized. In the rabbit, Kimmel (1944) could distinguish the glossopharyngeal and vagal ganglia by day E11 and found that afferents of the two ganglia entered the medulla in close proximity to the efferents of their nerves.

Results. The rudiments of the superior and inferior ganglia are recognizable by day E12 (Fig. 49) in the form of several cell clusters some distance laterally from the rhombencephalon. This chain extends from the hyoid arch, through aortic arches III and IV, as far as the visceral pouch, and they seem to be contiguous with the placodes of these regions (Fig. 49B). The components of this complex could not be identified with certainty at this age. In day E13 rats (Fig. 50) parts of the root ganglia were contiguous with the epithelium of the oral cavity (Fig. 50A) and parts of the trunk ganglia with the epithelium of the pharyngeal cavity (Fig. 50B). A fiber bundle was associated with the ganglia of IXGi and XGi, but its identity (IX–X, XI, or XII) could not be determined. A boundary cap was present in the region of the lower rhombencephalic alar

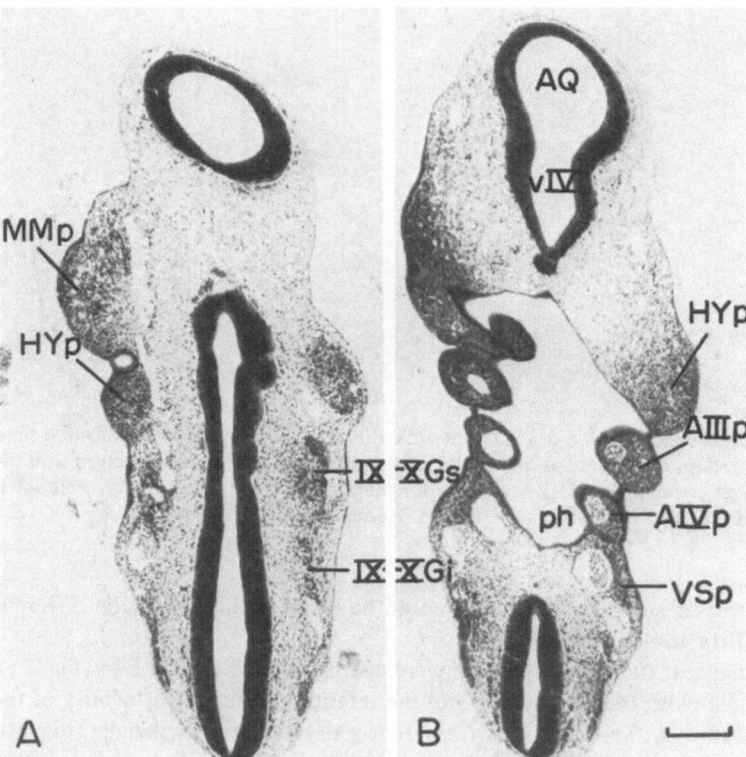

Fig. 49. *A* Horizontal section from a day E12 rat with the lateral chain of ganglia identified as the primordia of the glossopharyngeal and vagal ganglia. *B* A more ventral section to show the relation of the ganglia to the placodes of aortic arches III and IV, and the visceral arch. Paraffin. *Scale:* 200 μm

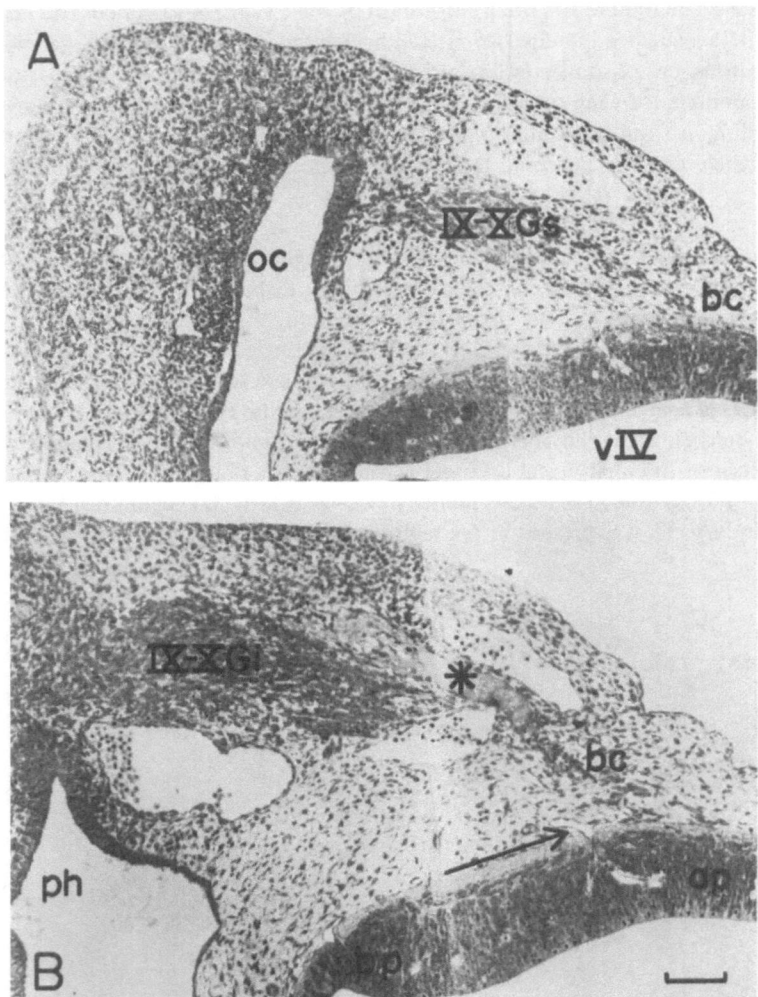

Fig. 50. *A* Horizontal section from a day E13 rat with the superior (root) ganglia, showing their relation to the epithelium of the lateral oral cavity. *B* A more ventral horizontal section with the inferior (trunk) ganglia and their relation to the epithelium of the pharyngeal cavity. *Asterisk* is over an unidentified (IX, X, XI, or XII?) nerve bundle. Methacrylate. *Scale:* 100 µm

plate (Fig. 50), and its spindle-shaped cells could be distinguished from the differentiating neurons of the root ganglia.

The four ganglia of the IX–X complex were distinguishable by day E14 (Fig. 51), the superior ganglia close to the medulla, and the inferior ganglia in the vicinity of the tongue. In day E14 rats, the neurons of the earlier-generated inferior ganglia (Fig. 48) were more advanced in their cytological maturation (Fig. 52B) than the neurons of the superior ganglia (Fig. 52A) that were produced mostly (Fig. 48) on this day. In day E14 rats, proximal fibers associated with these ganglia could be traced, by way of cell aggregates resembling the boundary cap, into the medulla, and distal fibers to the

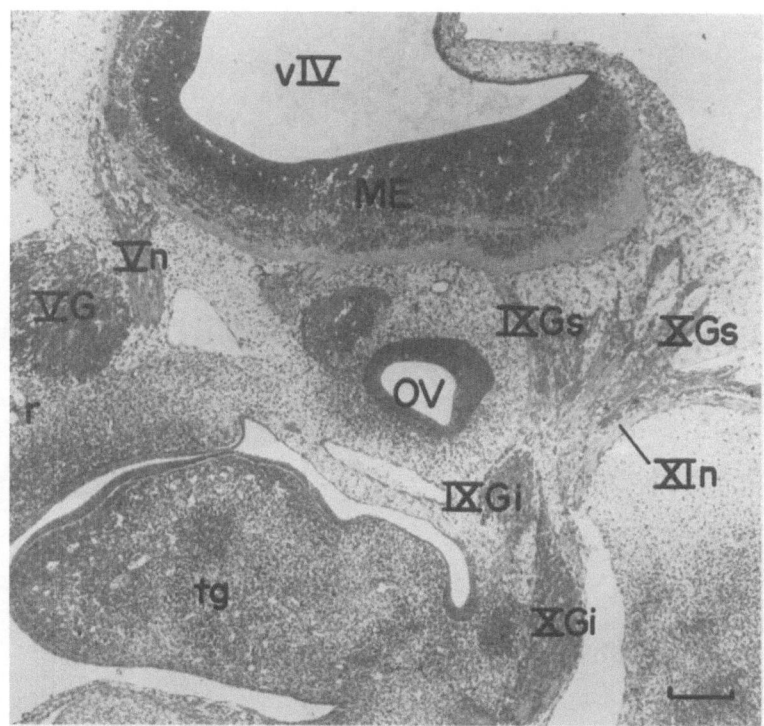

Fig. 51. Sagittal section through the head and tongue of a day E14 rat, showing the individual glossopharyngeal and vagal ganglia. Methacrylate. *Scale:* 200 μm

region of the tongue and beyond. But we could not determine the identity of these fibers. By day E15 the cytological differentiation of the neurons of the superior ganglia was comparable to that of the inferior ganglia (Fig. 53). A suggestion of the establishment of the gross circuitry of the IX–X ganglionic complex was the appearance of a large nerve in the depth of the tongue (Fig. 54).

Comments. In contrast to Adelmann's (1925) view that the ganglia of IX and X originate from a single postotic primordium, we were able to identify several discrete clusters of germinal cells as early as day E12. By day E13, the rudiments of the superior (root) ganglia could be identified in the vicinity of the hyoid arch, and the inferior (trunk) ganglia in relation to the aortic arches. In line with our thymidine-radiographic datings, we found that the cytological differentiation of neurons of the inferior ganglia antedated the differentiation of neurons of the superior ganglia. These observations reinforce the hypothesis that, from a morphogenetic point of view, the subdivision should not be between the ganglia of IX and X but between the root and the trunk ganglia. The embryonic location of the latter in the region of the tongue and other mucous components of the oropharyngeal region is reconcilable with the anatomical evidence that the petrosal and nodose ganglia innervate the tongue, pharynx, larynx, trachea, and esophagus, as well as the thoracic and abdominal viscera. The superior ganglia are supposed to provide cutaneous afferents to parts of the ear; but in view of

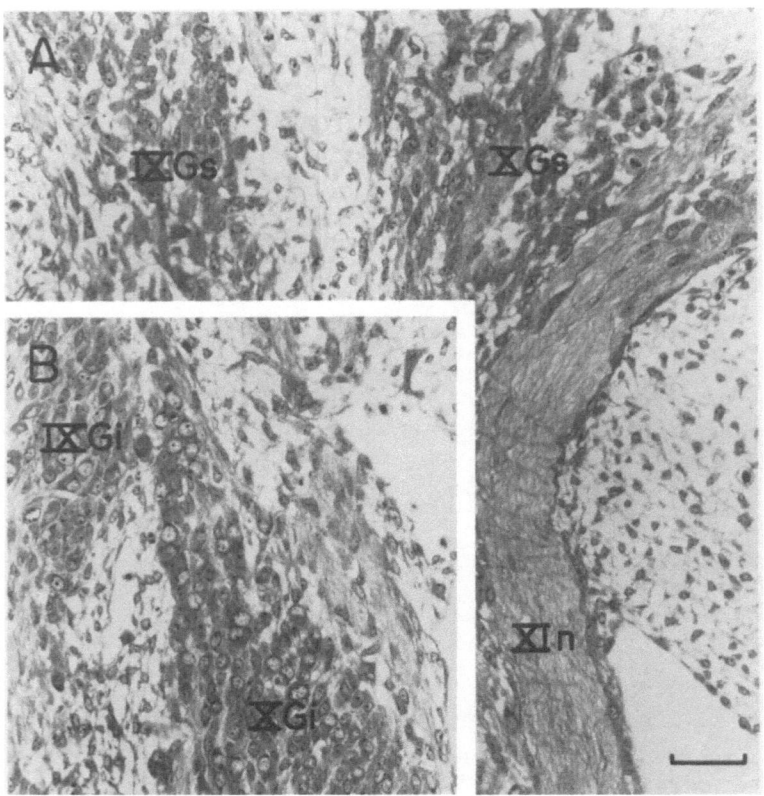

Fig. 52. *A* Sagittal section from a day E14 rat illustrating the primitive cells of the two superior ganglia. *B* The neurons of the earlier-produced two inferior ganglia show signs of cytological maturation. Methacrylate. *Scale:* 50 μm

the relatively large size of the superior ganglion and the jugular ganglion, one would expect a wider distribution.

6.3 Development of some Medullary Nuclei Associated with the Glossopharyngeal and Vagal Nerves

Background. In this phase of the study we have investigated the development of several medullary nuclei that are structurally and functionally related to nerves IX and X. Of these, the ambiguus nucleus is a horizontal column in the medullary reticular formation, composed of typical motor neurons. The nucleus is the source of vagal efferents to the striated muscles of the pharynx, larynx, and upper esophagus (Szabo and Dussardier 1964; Lawn 1966). Its neurons are produced relatively late, with a peak on day E15 (Altman and Bayer 1980a: Fig. 4A). Anterior to the ambiguus nucleus we have distinguished a column of similar cells that are generated very early, with a peak of production on day E11 (Altman and Bayer 1980b: Fig. 2): we identified this cytogenetically discrete structure as the retrofacial nucleus. According to Lawn (1966),

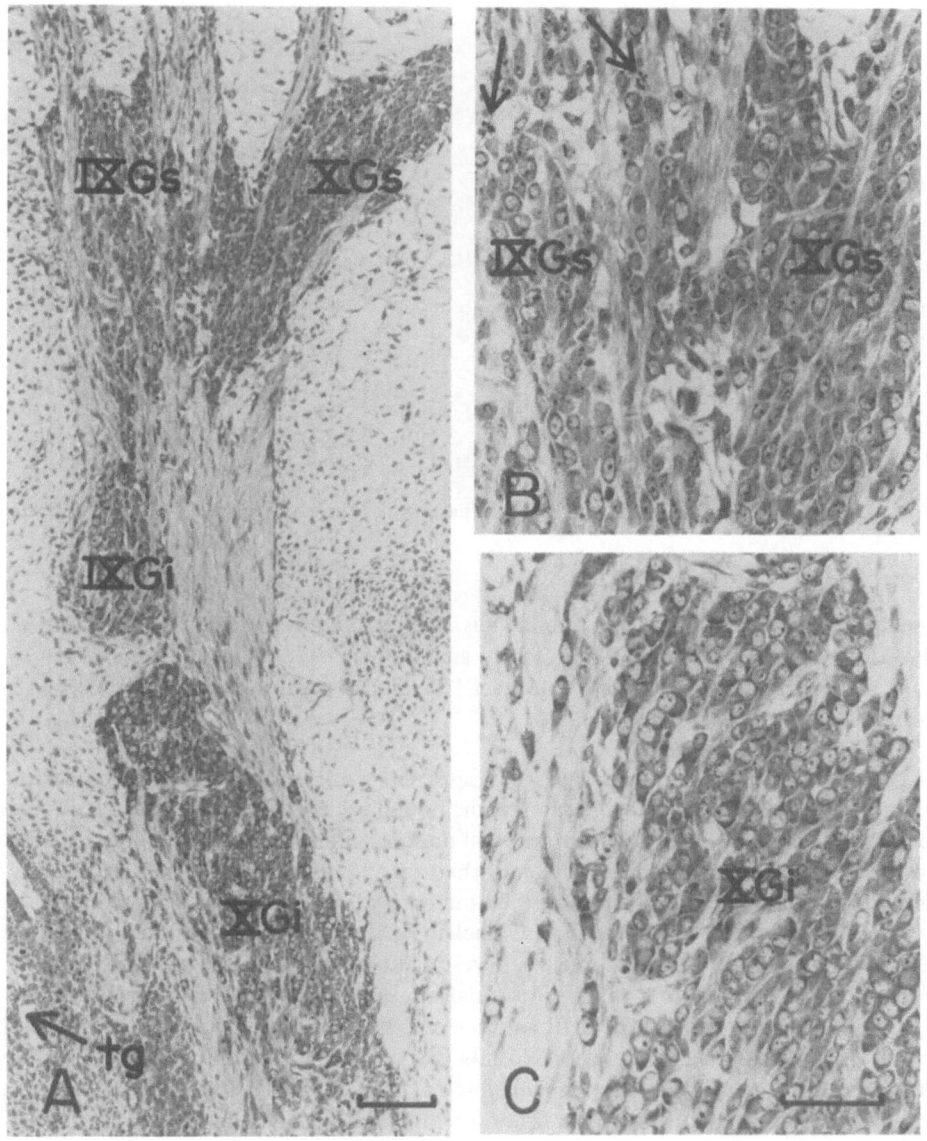

Fig. 53. *A* Sagittal section from a day E15 rat, showing the maturing neurons of the superior ganglia (detail in *B*) and the inferior ganglia (detail in *C*). In the superior ganglia a few mitotic cells are
still seen *(arrows)*. Methacrylate. *Scales: A*, 100 μm; *B, C*, 50 μm

this aggregate of motor neurons (which he described as the rostral portion of the ambiguus nucleus) is the source of glossopharyngeal efferents. The third structure that we
have briefly considered is the nucleus of the solitary tract. This nucleus is situated in
the dorsal medulla, extending from the level of the facial nucleus to the border region
of the spinal cord. The solitary nucleus is composed of small, densely packed cells that
receive afferents from nerves VII, IX, and X. The efferents are distributed to several

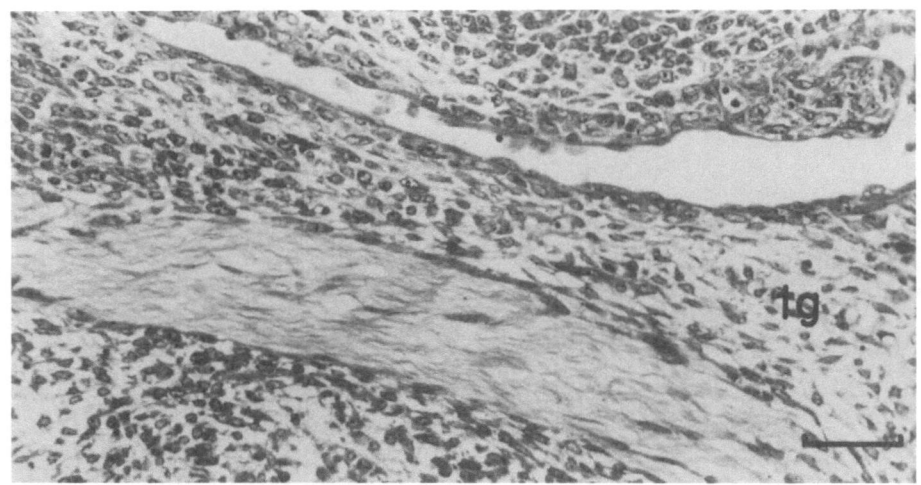

Fig. 54. Unidentified nerve (IX, X, or XII?) in the tongue of a day E15 rat. Methacrylate. *Scale:* 50 μm

medullary nuclei, including the ambiguus nucleus and retrofacial nucleus (Morest 1967; Cottle and Colorescu 1975; Loewy and Burton 1978). The neurons of the solitary nucleus are produced between days E12 and E15 with a peak on day E13 (Altman and Bayer 1980a: Fig. 4D).

Results. A stream of spindle-shaped cells, apparently migrating from the basal plate of the rhombencephalon in the direction of the boundary cap of IX—XGs was recognized on day E13 (Fig. 50B). This stream was still present on day E14 (Fig. 55A) but at least some of these cells reached the region of their target (Fig. 55B), and their axons, leaving the medulla, began to penetrate the boundary cap of IXGs (Fig. 55C). Since most of the motor neurons of the ambiguus nucleus are not generated until day E15, while the neurons of the retrofacial nucleus have originated by day E13, we identify these cells as the young neurons of the retrofacial nucleus (motor nucleus of the glossopharyngeal?). We were able to identify another stream of cells in the region of the basal plate of the upper medulla in day E16 rats (Fig. 56). It is possible that these are the late-generated neurons of the ambiguus nucleus; however, in the available material we could not trace this stream to the region of the vagal boundary cap. We were also unable to trace vagal or glossopharyngeal afferents to the solitary nucleus in young embryos. We identified the fibers of the solitary tract in day E17 rats (Fig. 57A—C); by this time the neurons of the ambiguus nucleus showed evident signs of cytological differentiation (Fig. 57D).

Comments. In the preceding studies we were able to identify the germinal sources and migratory paths of the trigeminal and facial motor neurons. We were handicapped in the present study insofar as the nuclear origins of the glossopharyngeal and vagal efferents could not be established with certainty. We were somewhat aided in our effort by the marked difference in the time of origin of neurons of the retrofacial nucleus and ambiguus nucleus. (The neurons of the retrofacial are the earliest-produced cells

62

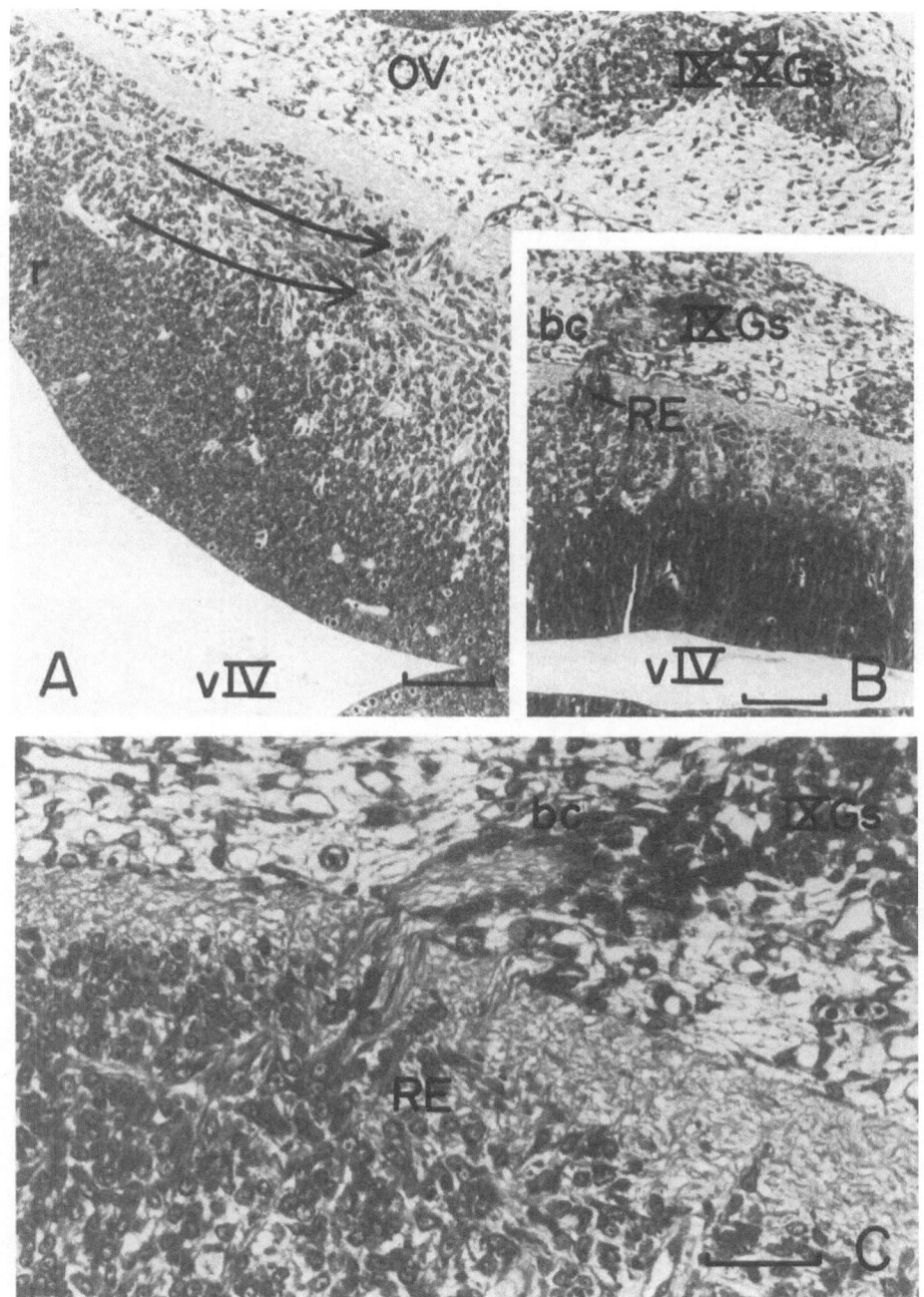

Fig. 55. *A* Horizontal section from a day E14 rat showing a stream of spindle-shaped cells, presumably neurons of the retrofacial nucleus, in apparent migration in the direction of the superior ganglia of IX and X. *B* In the same animal some of these cells appeared to reach their target region in the vicinity of the boundary cap of IXGs. *C* In another section fibers issuing from these cells are seen leaving the medulla and penetrating the boundary cap of IXGs. Methacrylate. *Scales: A, B,* 100 μm; *C,* 50 μm

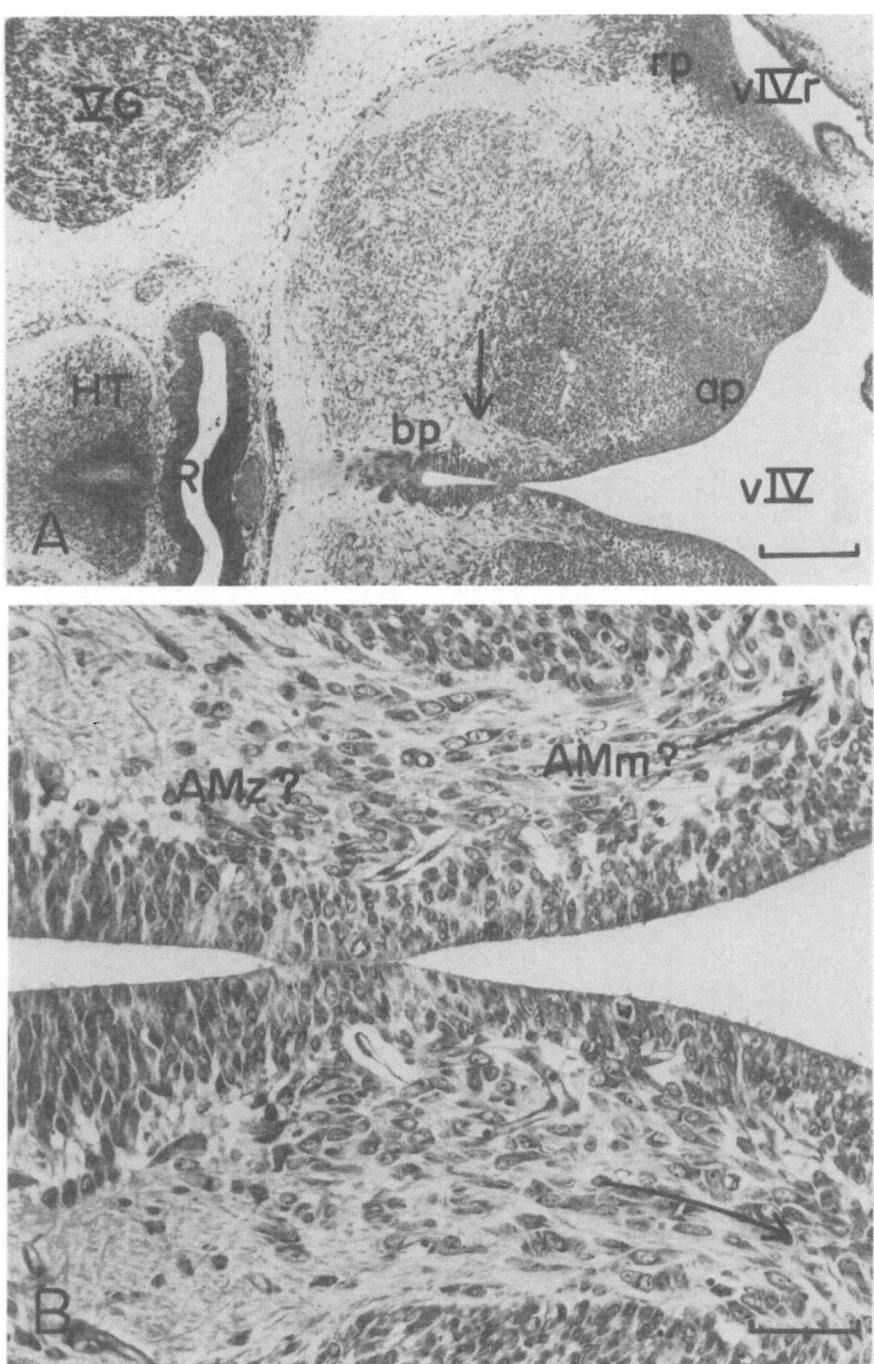

Fig. 56. *A* Horizontal section from a day E16 rat. *Arrow* points to a region of the basal plate which may be the source of a stream of spindle-shaped cells migrating in the caudal direction. *B* Detail of the basal plate region. On the basis of their late production, these cells may be the motor neurons of the ambiguus nucleus. Methacrylate. *Scale: A*, 300 μm; *B* 50 μm

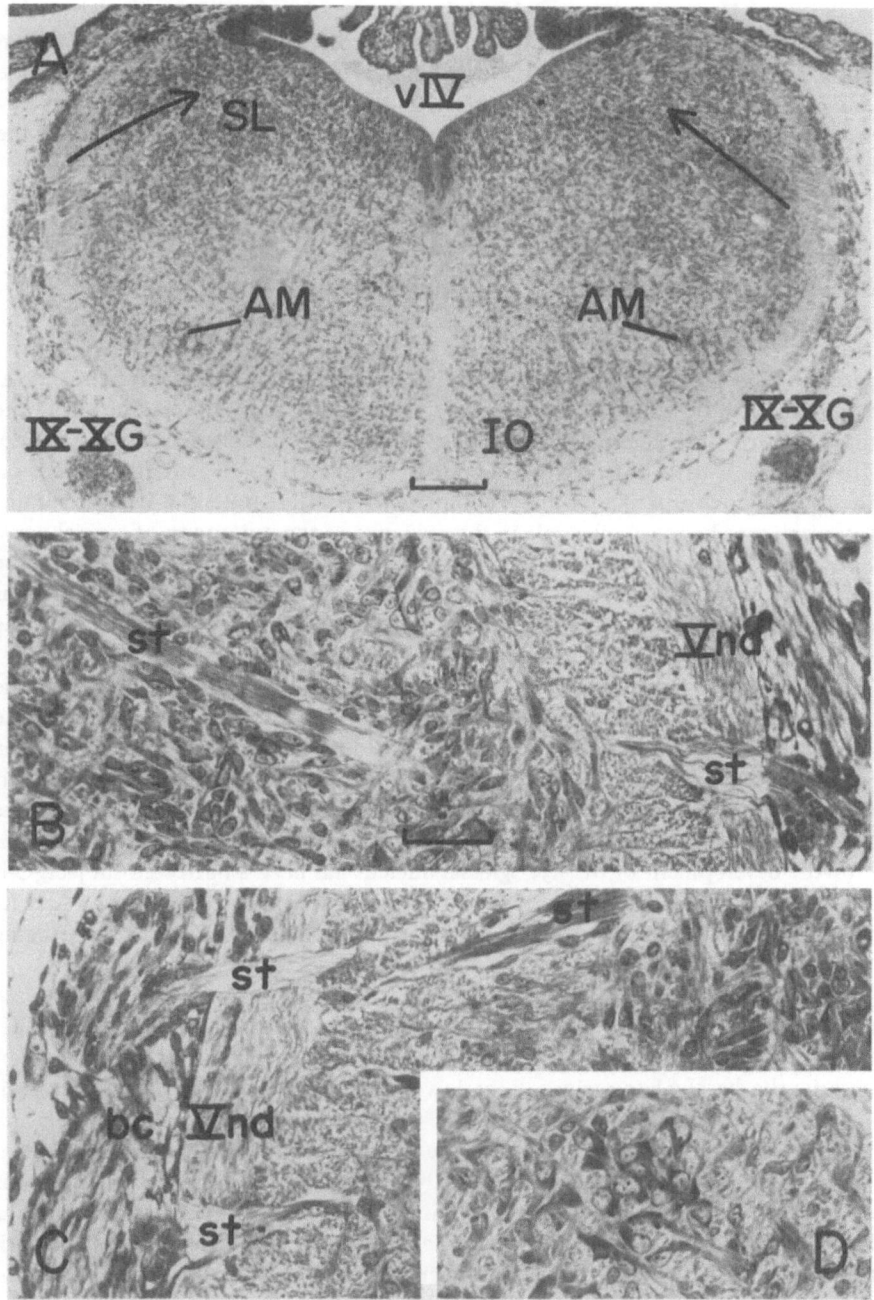

Fig. 57. *A* Coronal section of the medulla at the level of the inferior olive with fascicles of the solitary tract *(arrows)* coursing in the direction of the solitary nucleus. *B* Detail of the solitary tract from the right. *C* The same from the left. *D* The differentiating neurons of the ambiguus nucleus. Methacrylate. *Scales: A*, 200 μm; *B–D*, 50 μm

among the cranial nerve motor nuclei and the neurons of the ambiguus the latest; Altman and Bayer 1980a, b, e; 1981). We inferred that the cells of the migratory stream that begin to move in the direction of the boundary cap of IX and X on day E13 are the young retrofacial nucleus neurons, whereas the cells that migrate as late as day E16 from the basal plate of the rhombencephalon in the caudal direction could be the neurons of the ambiguus.

7 An Aspect of the Development of the Hypoglossal Nucleus and Nerve

Background. The observations made so far indicate that the afferents of the cranial nerve ganglia (V, VII, VIII, IX, and X) enter the medulla at specific points where the boundary caps of these nerves closely adhere to the medulla. These observations support the hypothesis that homonymous boundary caps exert some guiding influence on the directional growth of the proximal components of afferent nerves. Whether or not the boundary caps have a similar role in the guidance of the outgrowth of cranial nerve efferents is not so obvious, because in the case of mixed nerves (V, VII, IX, and X) the same effect could be achieved by a guiding influence exerted within the central nervous system by the penetrating afferents. Accordingly, we thought it worthwhile to include in this investigation a purely motor nerve of the medulla, the hypoglossal nerve.

The hypoglossal nucleus (nucleus of XII) forms a longitudinal column near the narrowing, posteroventral portion of the fourth ventricle, extending rostrally from the level of the prepositus nucleus to the region of the dorsal column nuclei caudally. The large motor neurons of XII are the source of efferents that leave the medulla in numerous rootlets ventrally, reach the inferior ganglion of X, and turn in the direction of the tongue, where they distribute to its intrinsic and extrinsic muscles (Brodal 1981). The study of the development of the hypoglossal nucleus has been a neglected subject. Our thymidine-radiographic datings (Altman and Bayer 1981a: Fig. 4A) showed that its neurons are produced between days E11 and E12, with over 80% of the cells being generated on day E12.

Results. The differentiating motor neurons of the hypoglossal nucleus were first recognized in day E14 rats (Fig. 58). These cells could be distinguished from all other elements in the basal plate region of the lower medulla by several criteria. The perikarya were the largest in this region of the differentiating zone (mantle layer), and they formed clusters that extended from the neuroepithelium and, curiously, spilled into the white matter (marginal layer) laterally. But the most conspicuous feature was the association of these cell clusters with fibers that left the medulla in numerous bundles (Fig. 58A) and penetrated the parallel clusters of primitive cells located peripherally (Fig. 58B). We identify this peripheral chain of cells strung along the lower medulla as the hypoglossal boundary cap. In the regions where fiber bundles left the medulla, tubular formations were recognized (Fig. 59) in which the motor neurons "slid" as far as the pial border (Fig. 58B), and where boundary cap cells appeared to be "pulled" to the vicinity of the parenchyma (Figs. 58B, 59).

By day E15 the neuroepithelium of the basal plate of the lower medulla shrank considerably (Fig. 60A). As a result of this, and due to the active migration of cells

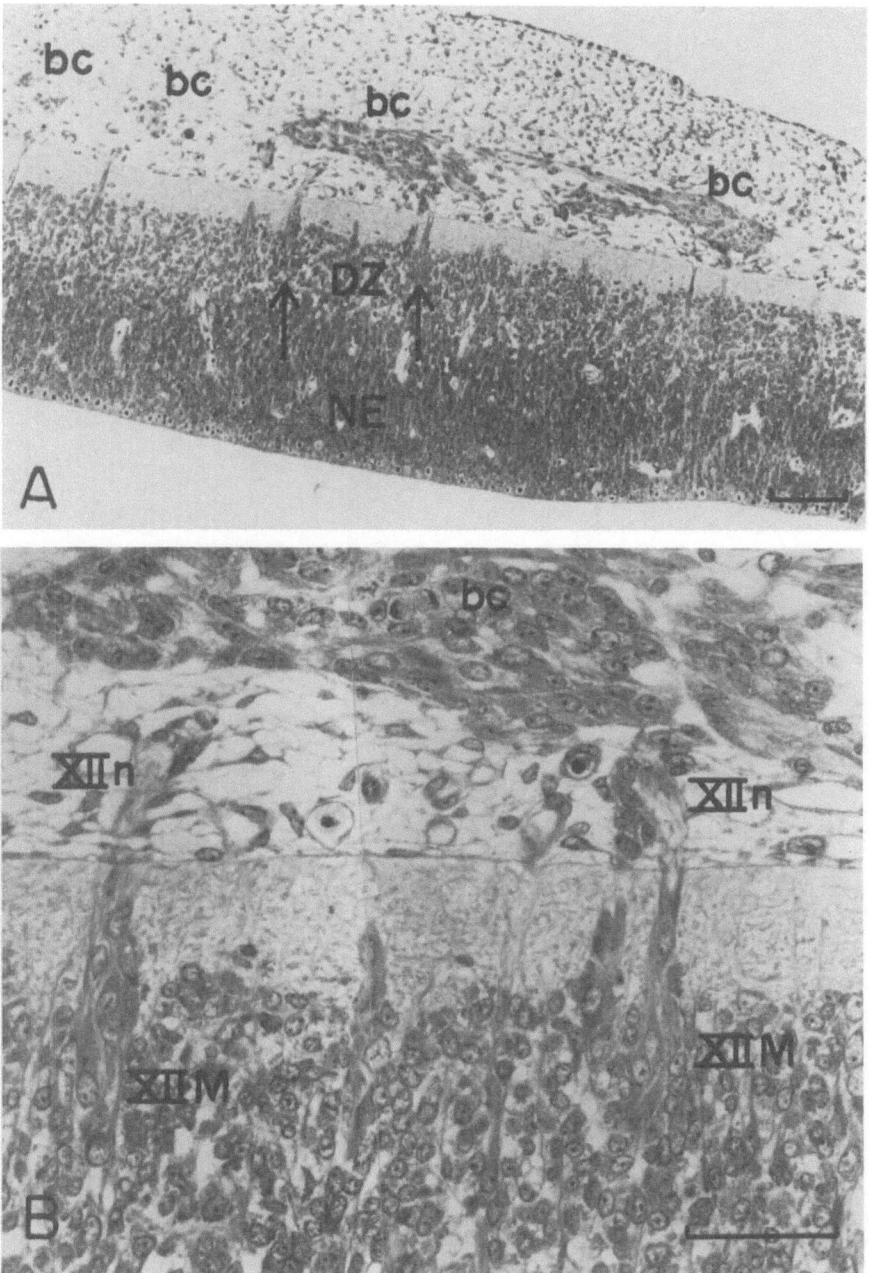

Fig. 58. *A* Horizontal section of the lower medulla in a day E14 rat with clusters of peripheral cells, designated as the boundary cap of XII, and corresponding clusters of central neurons *(arrows)* considered to be the motor neurons of XII. *B* Detail showing clusters of the boundary cap and of motor neurons of XII, and efferents between the two. Methacrylate. *Scales: A*, 100 μm; *B*, 50 μm

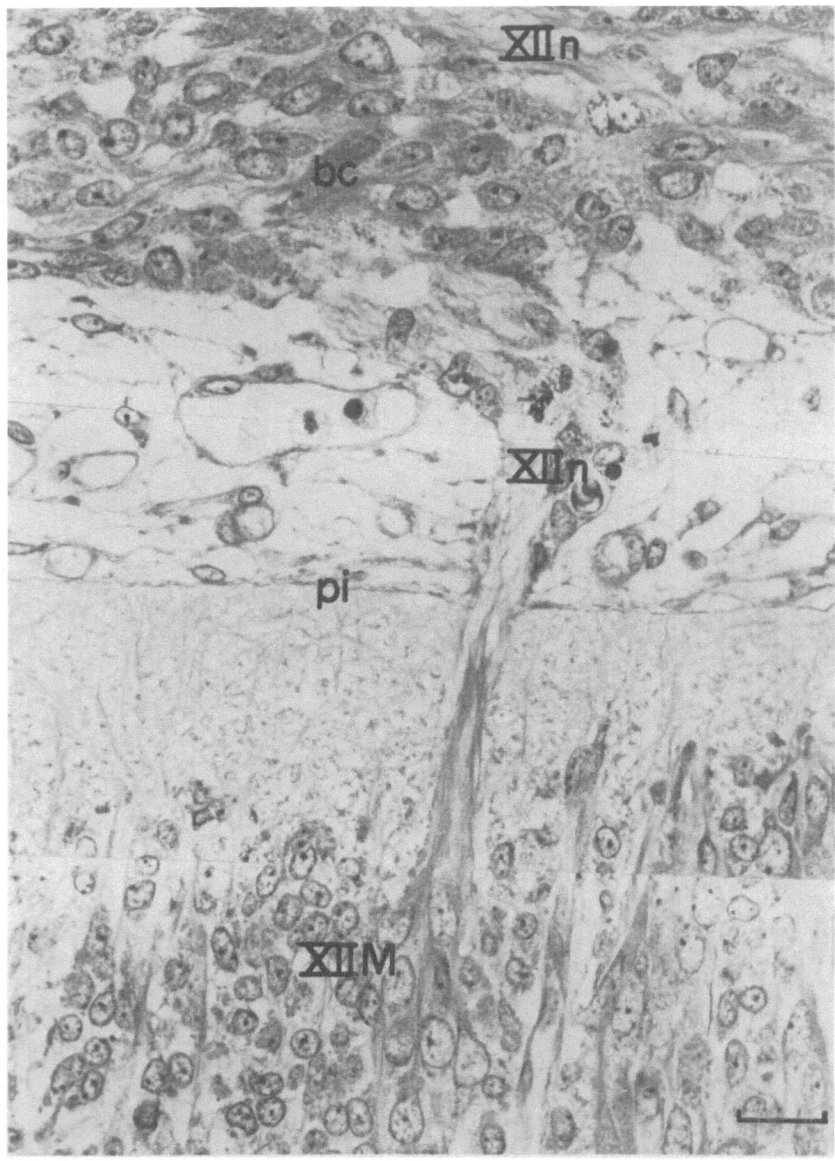

Fig. 59. The boundary cap cells and motor neurons of XII at higher magnification. Note tubular structures where the pial membrane is absent and efferents approach the cells of the boundary cap that are near the pial membrane. *Scale:* 20 μm

from the neuroepithelium of the alar plate into the ventral half of the medulla (Fig. 60B), the juxtaventricular hypoglossal motor neurons now appeared to be situated in their typical medial position. By day E16, the initial clustering of motor neurons was no longer in evidence and the nucleus formed a continuous longitudinal column.

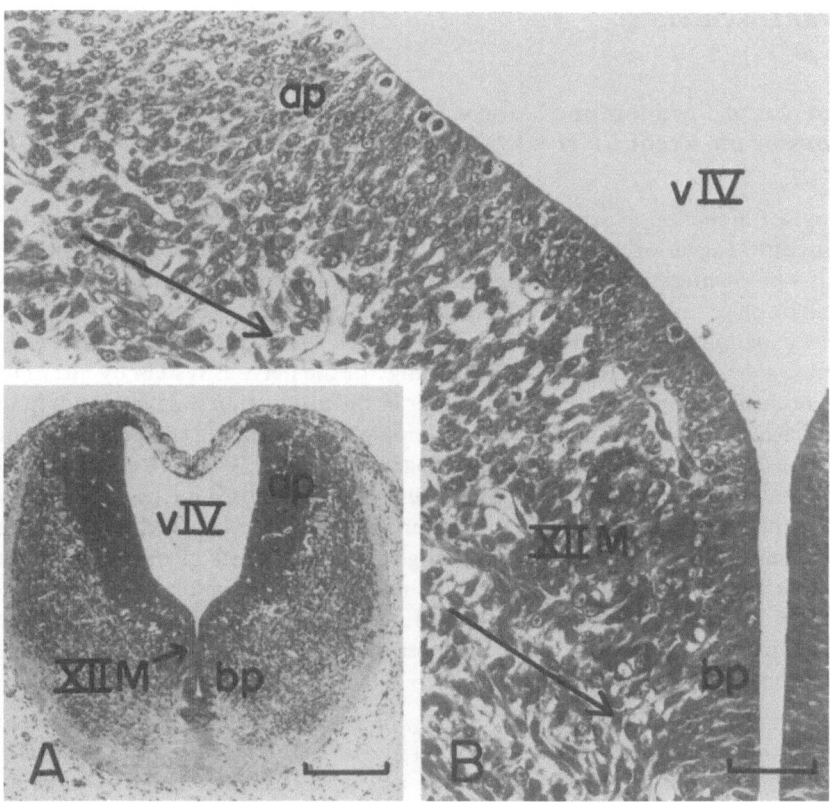

Fig. 60. *A* Coronal section through the lower medulla in a day E15 rat. *B* Detail to show the differentiating motor neurons of XII adjacent to the receding neuroepithelium of the basal plate. *Arrows* point to cells apparently migrating from the alar plate region and settling laterally. *Scales: A*, 300 μm; *B* 50 μm

Comments. The observations made in this phase of our investigation indicate that peripheral boundary caps in the medulla are associated not only with sensory and mixed cranial nerves but also with a purely motor nerve. In adults, the fibers of the hypoglossal nucleus leave the medulla in several rootlets, and in young embryos we could identify corresponding efferent bundles that issued from clusters of motor neurons of the lower medulla and grew in the direction of corresponding aggregates of cells peripherally. It appears, therefore, that boundary caps are not derivatives of cranial nerve ganglia and that they may play a direct role not only in the guidance of the growth of afferents from the cranial nerve ganglia centrally, but also in the guidance of efferents into the periphery.

8 General Discussion

8.1 Sequence of Production of Motor Neurons, Ganglion Cells, and Sensory Neurons of the Cranial Nerves

The concept of induction (Spemann 1938) implies that during morphogenesis earlier-arising structures trigger or exert an organizing influence on later-emerging structures. The idea has been entertained by developmental neurobiologists for some time that induction might play a role in the development of the circuitry of the nervous system, and various experimental procedures have been devised to test this hypothesis (reviewed in Jacobson 1978; pp. 253–307). The "wiring" of the nervous system by means of inductive mechanisms presupposes a fixed order in the production of its elements, and this requirement is satisfied by the precise chronological sequence in the production of neurons, as revealed by recent thymidine-radiographic studies. With information now at our disposal about the time of origin of neurons of the motor nuclei, the sensory ganglia, and the sensory nuclei of the cranial nerves, it appeared worthwhile to inquire if there is a uniform order in the sequence of generation of these elements such that we may entertain the possibility of inductive mechanisms operating in the morphogenesis of the medullary circuits of the cranial nerves. We will first consider possible neuronal interactions, then deal with the possible role of a nonneuronal guidance system, the cranial nerve boundary caps.

Figure 61 summarizes the relevant neurogenetic datings obtained in the present investigation and in an earlier series of studies (Altman and Bayer 1980a–d, 1981). In the trigeminal-facial system (first column), the neurons of ganglia V and VII are produced before their target neurons in the sensory nuclei V (including the neurons of Vs shown in the bottom of the third column). Since the ganglion cells in this system are produced before the neurons of the sensory nuclei, peripheral neurogenesis must be independent of central events. On the other hand, an inductive influence exerted by the neurons of the cranial nerve ganglia on the neurons of the sensory nuclei would be chronologically possible. In view of the appreciable distances involved, it is likely that such an influence would be mediated by way of afferents entering the medulla. As our observations indicated (Fig. 10), trigeminal afferents reach the medulla by day E13, a day before the peak production of the neurons of the principal sensory nucleus of V. We also observed the migration of cells (presumably of the principal sensory nucleus) in the direction of the entering afferents of V (Fig. 16B). It is also evident from the first column in Fig. 61 that the generation of the motor neurons of V and VII must be independent not only of the sensory nuclei of V but also of their sensory ganglia, because the motor neurons of V are produced before the ganglion cells of V. We did entertain the possibility that the migration of the motor neurons of V in the direction of the entering afferents of V is guided by the latter, but the generality of such an effect is ruled out by the fact that the motor neurons of VII do not migrate in the direction of the entering nerve of VII. In summary, the established chronological relations suggest that in the trigeminal-facial cranial nerve system, neurogenesis of the sensory ganglia and motor nuclei must be independent of each other and of the sensory nuclei, but that the sensory ganglia could exert an influence, by way of afferents, on the migration (and less likely on the proliferation) of neurons of the sensory nuclei.

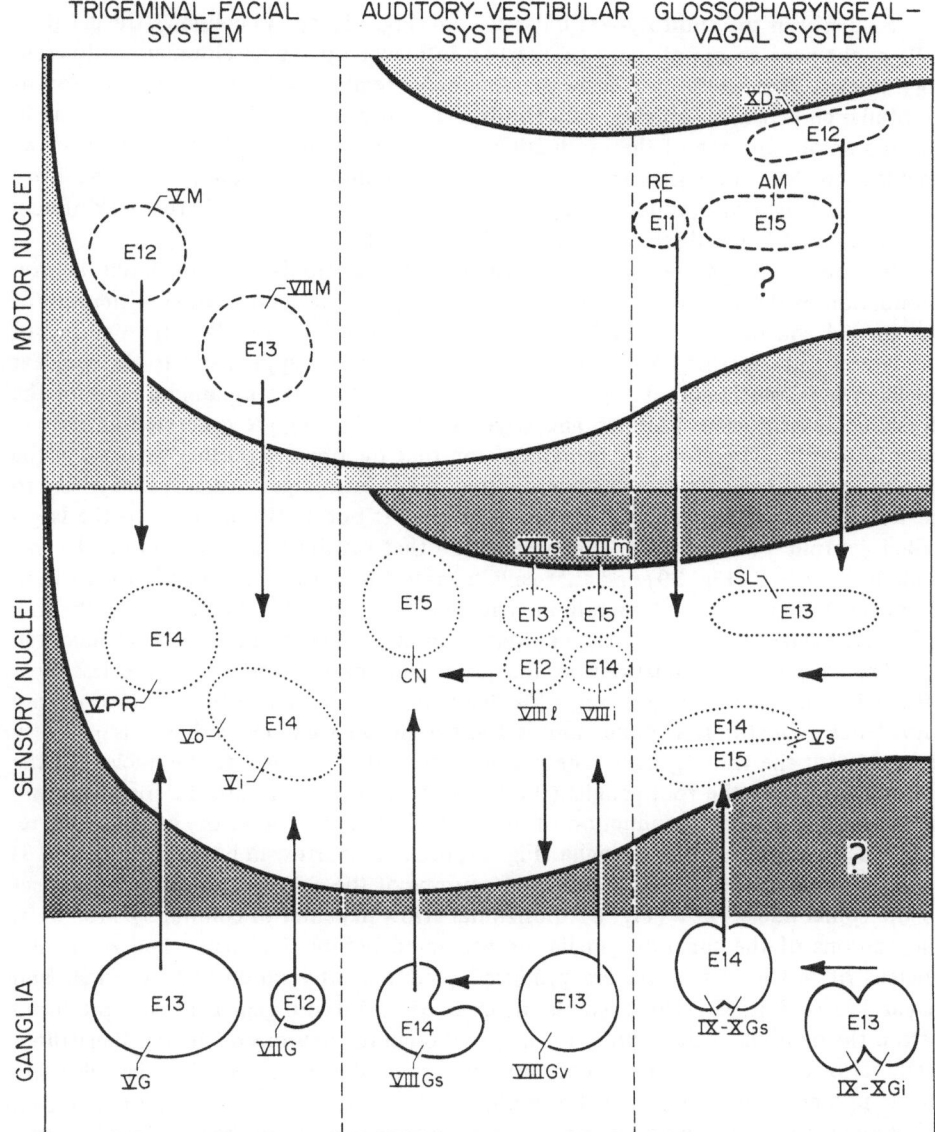

Fig. 61. The temporal order of neurogenesis in the motor nuclei *(first row)*, sensory nuclei *(second row)*, and sensory ganglia *(third row)* in three related cranial nerve systems: the trigeminal-facial *(first column)*, the auditory-vestibular *(second column;* this system has no motor nuclei), and the glossopharyngeal-vagal *(third column)*. *Embryonic ages* refer to days with peak neurogenesis. *Vertical arrows* refer to cytogenetic gradients (from earlier to later) in the motor nuclei and the sensory ganglia in relation to the sensory nuclei. *Horizontal arrows* indicate the matching cytogenetic gradients between the sensory ganglia and between the sensory nuclei in the auditory-vestibular and the glossopharyngeal-vagal systems

Similarly in the auditory-vestibular system (Fig. 61; second column), the ganglion cells might have an opportunity to exert an influence on the development of the sensory nuclei. First, there is a corresponding order (early-to-late) in the production of vestibular and cochlear ganglion cells (horizontal arrow in bottom of column 2), and in the temporal sequence of the production of neurons of the vestibular nuclei and the cochlear nuclei (horizontal arrow in center of column 2). Second, in the auditory system the height of production of ganglion cells antedates (day E14) the peak of neuron production in the three cochlear nuclei (day E15).

The relationships are more complicated in the vestibular system. While neuron production in the vestibular ganglia precedes the peak production time of neurons of the medial and inferior vestibular nuclei, this does not hold for the large neurons of the lateral and superior vestibular nuclei. Importantly, this applies also to the cochlear nuclei, where the large neurons, for instance the cells of the pyramidal layer of the dorsal cochlear nucleus (Altman and Bayer 1980c: Fig. 21), are generated as early as days E12—E14. Therefore, the generalization that the primary sensory neurons of the cranial nerve ganglia are produced before their medullary target structures may have to be restricted to their small-celled (input?) elements but need not apply to the large-celled (output?) components. Our observation that vestibular afferents enter the medulla by day E13 (Fig. 39) and that cells from the alar plate (presumably the earliest-produced Deiters neurons) migrate in that direction allows for the possibility that the influence exerted by the primary neurons is on cell migration not on cell production.

The chronological relations in the glossopharyngeal-vagal system are less clear (Fig. 61, third column), and our difficulty here is compounded by a lack of precise knowledge of anatomical connections in this system. As we noted earlier, it is presently believed that the trunk ganglia (the petrosal and nodose) project to the nucleus of the solitary tract, and the root ganglia (the superior and the jugular) to the spinal trigeminal nuclei. The earlier production of neurons of the trunk (inferior) ganglia with respect to the root (superior) ganglia (Fig. 61, horizontal arrow in bottom of column 3) corresponds to the earlier production of neurons of the solitary nucleus with respect to the spinal nuclei of V (Fig. 61, horizontal arrow in center of column 3). Moreover, the neurons of the superior ganglia are generated before the small-celled zonal subnucleus of V whose neurons are produced predominantly on day E15 (Altman and Bayer 1980a: Fig. 4E). However, the neurons of the inferior ganglia are not produced before the neurons of the solitary nucleus (although the neurons of the facial ganglion, which also send fibers to the solitary nucleus, are produced earlier). Even less clear are the temporal relations between the motor nuclei of IX—X and their sensory nuclei. While the neurons of the dorsal nucleus of X, and the neurons of the retrofacial nucleus (neurons of IX?) are generated before the neurons of the solitary nucleus and the spinal nuclei of V (corresponding to the temporal relationship observed in the trigeminal-facial system), the neurons of the ambiguus nucleus are produced after most components of the sensory nuclei of IX—X. We were also handicapped in the examination of this system in that we were unable to determine the time of arrival of afferents, and we were uncertain about the identity of the two migratory streams that we observed in day E14 (Fig. 55A) and day E16 (Fig. 56) rats.

Accordingly, with the exceptions noted, we suggest that the chronological relationships obtained allow for the possibility that an organizing influence is exerted by the relatively early-generated ganglion cells, by way of their afferents, on the migration of sensory neurons in the medulla to their final settling sites. An influence of ganglion cells on the proliferation of neurons of the sensory nuclei is less likely, and any early influence on the motor neurons by either the ganglion cells or the sensory neurons is ruled out by the obtained chronological relations.

8.2 Placodal Origin of Ganglion Cells

Placodes are regional thickenings of the epithelium which, on the basis of their distribution in the head region, are believed to be directly related to the special and the cutaneous head receptors and possibly to certain visceral elements. The possibility that placodes are the source of both sensory cells and ganglion cells is best illustrated in the case of the otic vesicle. It is well known that this vesicle begins with the invagination of an epithelial thickening which detaches from the surface, when the otic cyst is closed (Figs. 23, 24). After formation of the otic vesicle two waves of cells leave the external wall of the vesicle, the first coincident with the time of origin of vestibular ganglion cells (Figs. 36, 37), and the second with the time of origin of spiral ganglion cells (Figs. 41, 42). During the next phase of development (which was not described in this study), the bulk of the otic vesicle differentiates as the membranous labyrinth and the membranous cochlea with their hair cells and accessory elements.

Our observations are compatible with the assumption that all the cranial nerve ganglia derive from placodes. We have not attempted here to delineate precisely the location and extent of the different placodes relative to the branchial arches. The placodes are usually described as discrete cell clusters rather than a continuous cellular sheet, and two observations support this. Our quantitative data indicated that the chronological pattern of neurogenesis is quite different in two such related ganglia as IXi and IXs, and Xi and Xs (Fig. 48). The pattern that we obtained suggests that the inferior ganglia derive from one placode with a similar time course of neurogenesis and the superior placode with a similar time course of neurogenesis and the superior ganglia from another placode with another pattern. The other relevant observation is summarized in Fig. 61 (bottom row). There is apparently no unified cytogenetic order in the production of sensory ganglia in the rostrocaudal direction but rather, in terms of peaks of cell production, a discontinuous sequence (E13, E12, E14, E13, E14, and E13), as if many of these ganglia had their independent life histories.

We have referred earlier to the widespread belief that the neural crest is either the exclusive or a contributory source of ganglion cells. Our results suggest that in the cranial nerves of the medulla, the neural crest is more likely to be the source of a non-neuronal transient embryonic system, the boundary caps.

8.3 Nerve-Specific Boundary Caps and Their Possible Role in the Guidance of Afferents and Efferents

In agreement with previous observations we have recognized a medial and a lateral component in the region that is usually considered to be the anlage of the trigeminal ganglion (Figs. 6, 7). We have referred to the hypotheses that these two primordia are either separate sources of the two components of the mature trigeminal ganglion, or of the presumably separate populations of its small and large neurons. However, our observations indicated that the lateral anlage is the sole source of the trigeminal ganglion itself, and the medial anlage of a nonneuronal structure, the trigeminal boundary cap. Reinforcing this interpretation was our subsequent finding that boundary caps are associated during development with all the cranial nerves of the medulla; including those which clearly do not have subdivisions (the facial ganglion), those in which a separate population of small and large neurons is not evident (VII, IX–X), and a pure motor nerve (XII) that is not associated with a sensory ganglion at all.

Our observations indicated that boundary caps are situated at the junctional region where the cranial nerves pierce the wall of the medulla. Our observations also suggested that the boundary caps are a source of primitive cells that spread along the nerves peripherally (Fig. 11). From these characteristics we inferred that the primitive cells of the boundary caps are the source of the cranial nerve Schwann cells, and possibly of the satellite cells of the ganglia of the sensory and mixed cranial nerves. Both the Schwann cells and the satellite cells were labeled in thymidine radiograms in animals injected as late as days E17+18, indicating that they are late-produced elements. In agreement with this, we could still recognized the boundary caps in association with several nerves as late as day E21. The boundary caps may disappear sometime during the perinatal period, as in a day P5 rat we could no longer find it in the trigeminal nerve.

The embryonic origin of Schwann cells has long been debated, centering around the question of whether they are derivatives of the neural crest, the neural tube, or both (Hörstadius 1950; Weston 1970). The experimental evidence seems to favor the view that the major source of Schwann cells is the neural crest (e.g., Harrison 1924; Detwiler and Kehoe 1939; Yntema 1943; Johnston 1966). We have not focused on this issue, but an incidental observation argues against the neural tube origin of Schwann cells; namely, that in young embryos (e.g., Fig. 11) cells are virtually absent in the central part of the cranial nerves but are densely packed peripherally. Typically, there was a sharp demarcation line at the central border of boundary cap as if these cells associated with the peripheral portion of the cranial nerves and, originating peripherally, were denied entry into the central nervous system. Insofar as boundary caps were present adjacent to the medulla in several cranial nerves during early phases of embryonic development, when the rudiments of the ganglia were still contiguous with placodes more laterally, our observations tend to support the idea of the neural crest origin of Schwann cells. With regard to the satellite cells, we observed mitotic cells in the trigeminal ganglion after the age of cessation of neurogenesis (Fig. 9B), suggesting the possibility of local proliferation of these elements. But this was not a common feature in most ganglia, where mitotic cells were usually absent after the time that neurogenesis has come to an end in terms of thymidine-radiographic datings. It is possible that the satellite cells are also derivatives of the boundary caps.

Figure 62 summarizes our observations regarding the sites where fibers of the cranial nerve ganglia enter the medulla. In all instances the afferents traverse boundary caps, irrespective whether this cell aggregate is located adjacent to the ganglion (VG, VIIG, VIIIGv, IX–XGs) or a short distance from it (VIIIGs, IX–XGi). Figure 63 illustrates where fibers of the cranial nerve motor neurons leave the medulla. The efferents, too, traverse boundary caps, irrespective of whether the motor perikarya (during embryonic development) settle close to the boundary cap (VM), a short distance from it (XIIM), or relatively far away (VIIM, RE, AM). These observations suggest that boundary caps may be "nerve-specific" guidance agents that determine where afferents and/or efferents establish contact between center and periphery, and what groups of fibers will constitute a specific cranial nerve. Relevant in this context is the

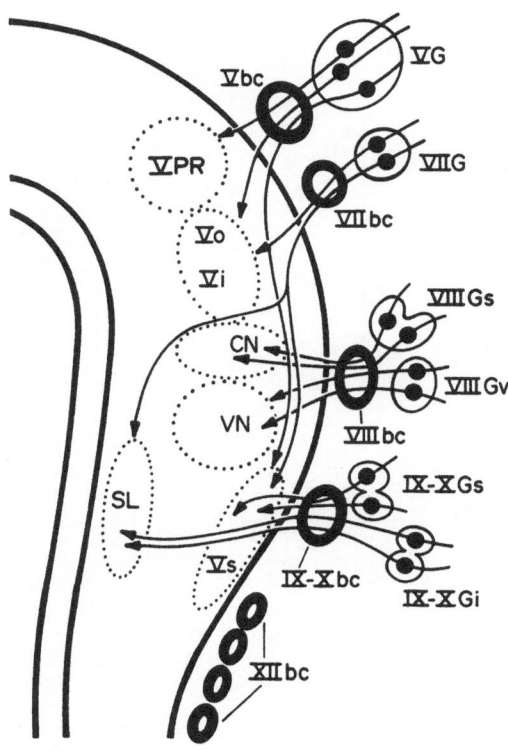

Fig. 62. The relation of the nerve-specific boundary caps to the site of entry of afferents of the sensory and mixed cranial nerves

observation (Yntema 1943) that the normal growth of motor fibers into the limb of *Ambystoma* is interfered with as a result of extirpation of the presumed source of Schwann cells. In contrast, the boundary caps seems to play little or no role (Fig. 62) in guiding the entering afferents to their targets centrally. Nor are they likely to be involved in specifying the course of the distal branches of afferents and efferents peripherally. The observations summarized in Fig. 64 also indicate that the boundary caps do not specify the migratory course of motor neurons originating in the medullary basal plate. While the trigeminal motor neurons do migrate in the direction of the trigeminal boundary cap, the facial motor neurons migrate in a different direction, the motor neurons of the retrofacial and ambiguous nuclei stop in the reticular formation, and the motor neurons of the hypoglossal nerve do not migrate at all.

8.4 Sites of Neuron Production, Routes of Migration, and the Relation Between Cytogenetic Zones and Neuroepithelial Zones in the Medulla

The somatic cranial nerve motor nuclei (the oculomotor, trochlear, abducent, and hypoglossal) are situated in the vicinity of the ventricular wall, whereas the branchial motor nuclei (the trigeminal, facial, retrofacial, and ambiguus) are situated some distance

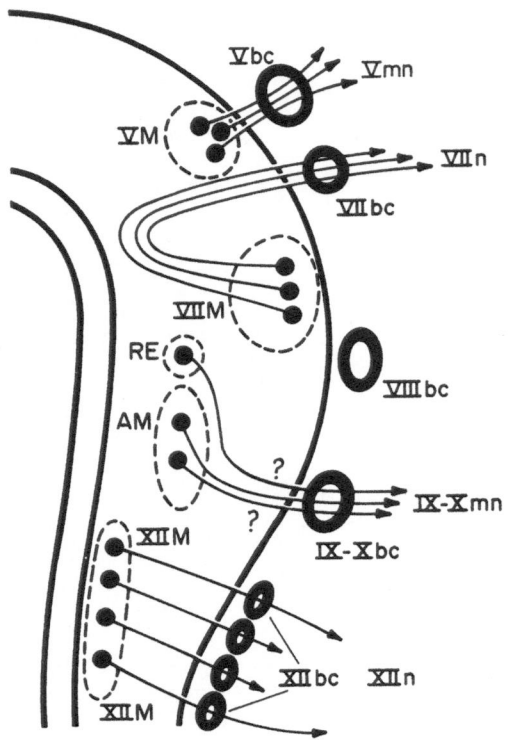

Fig. 63. The relation of the nerve-specific boundary caps to the sites of exit of the efferents of the motor and mixed cranial nerves

from it. In spite of this difference in final location, it has been maintained for some time (e.g., Bok 1915; Tello 1922; Windle 1933) that both classes of motor nuclei derive from a continuous, medially situated germinal matrix. This conclusion has received support from our observations summarized in Fig. 64. Apparently, neurons of the somatic motor nuclei (which, incidentally, supply fibers to pure motor nerves) settle near their site of production in the basal plate while the neurons of the branchial motor nuclei (which supply mixed nerves) migrate. We have been able to date and trace the migration of trigeminal and facial motor neurons, and (with lesser assurance) the motor neurons of the retrofacial and ambiguus nuclei.

According to Ariëns Kappers et al. (1936, vol. 1, p. 519), the migrating branchial motor neurons exemplify the operation of the principle of neurobiotaxis (Ariëns Kappers 1917). For instance, the neurons of the facial nucleus are presumed to migrate in the direction of medullary structures (not clearly specified) that are sources of major input to the facial neurons. Ariëns Kappers et al. went on to suggest that species differences in the location of the facial nucleus might be accounted for by differences in its anatomical connections in the different species. Unfortunately, it has never been satisfactorily specified what these central structures may be that induce the migration of the different branchial motor neurons to different sites. Topographic considerations would suggest that, in mammals, the trigeminal motor neurons migrate toward the entering afferents of V, the facial motor neurons toward the descending nerve of V,

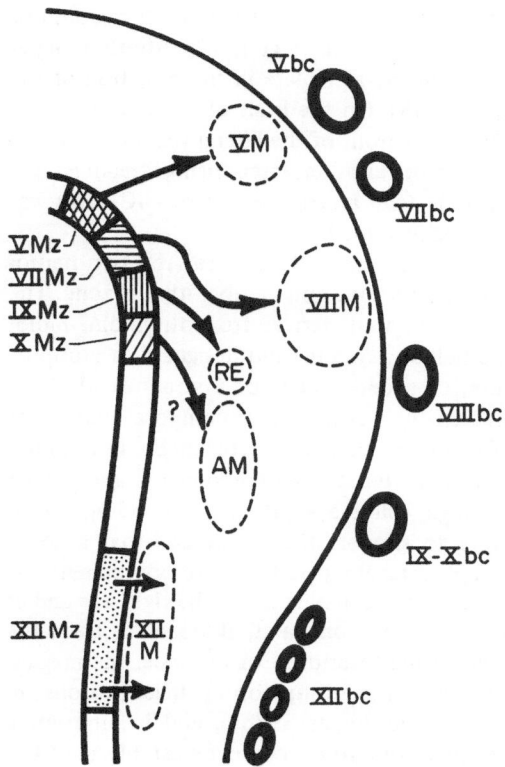

Fig. 64. The route of migration *(arrows)* of the cranial nerve motor neurons of the medulla from their sites of production (segments in the midline neuroepithelium) to their settling sites. The settling sites of the motor neurons seem to bear no obvious relationship to the location of the boundary caps

and the retrofacial and ambiguus neurons toward the fibers of the solitary tract. But this vague inference is not supported at present by either developmental, comparative, or experimental evidence. By inference, the neurons of the somatic motor nuclei do not migrate (or migrate only a short distance), because their input source is located close to where they are produced. This certainly holds for the oculomotor, trochlear, and abducent nuclei, which have intimate connections with the medially situated medial longitudinal fasciculus, an early-forming fiber tract. It is not clear at present what medially situated fiber tract arrests the migration of the hypoglossal neurons. In summary, these and previous considerations suggest that the central course of the growing axons of medullary efferents is specified by the boundary caps, and that the migration of their perikarya (possibly the growth and streaming of the dendrites) is guided by as yet unspecified central structures.

Not only motor neurons but also sensory neurons display migration. It was possible to trace the migration of neurons from the alar plate or rhombomere 1 in the direction of the entering afferents of V, and we suggested on the basis of our datings that these are the neurons of the principal nucleus of V. Because this migration toward the entering afferents of V occurs after the motor neurons have settled in this region, this new wave of cells displaces the motor neurons of V medially (Fig. 17). Another migratory stream was identified more caudally from the alar plate in the direction of the entering afferents of the vestibular nerve (Fig. 38). On the basis of our datings it was suggested that these cells are the young Deiters neurons of the lateral vestibular nuclei. This migratory stream could be distinguished from another, later-arising stream issuing from the lateral recess of the fourth ventricle (the recess plate; Fig. 44C); this was identified with the principal neurons of the cochlear nuclei.

We have recently introduced (Altman and Bayer 1980a, c, 1981b) two neuroembryological concepts, referred to as cytogenetic zone and neuroepithelial zone. The hypothetical construct of a cytogenetic zone or unit derives from thymidine-radiographic studies in the maturing or mature nervous system, and designates a group of related brain structures, a single structure, or a component of a structure that are distinguished from others in the same region by a shared spatiotemporal pattern of neurogenesis. The designation of a set of brain structures as a cytogenetic unit implies that by virtue of their shared cytogenetic characteristics its components are derived from a shared germinal source, perhaps a single cell-line, or the corresponding neuroepithelial zone or unit. Therefore, in order to support the hypothesis that a set of brain structures with similar autoradiographic labeling patterns are components of a single cytogenetic unit, it is necessary to identify in embryos (of the right age and at the right place) the corresponding neuroepithelial zone and, if possible, trace the migration of the cells from this site to their final location. For instance, the magnocellular neurons of the paraventricular, intranuclear, and supraoptic nuclei neurons are produced over a similar time span (Altman and Bayer 1978a), and in embryonic material their neuroepithelial source and migratory route could be established (Altman and Bayer 1978b). The two cytogenetic zones of the medulla that are relevant to the present discussion are zones MB and VE.

Cytogenetic zone MB (Altman and Bayer 1980b: Fig. 15) referred to the branchial motor nuclei of the trigeminal and facial nerves and the ambiguus nucleus. The nuclei were postulated to derive from a single neuroepithelial source because their neurons are produced in a sequential order, that is, showed a continuous cytogenetic gradient from rostral to caudal. We did not include the retrofacial nucleus as part of zone MB because its neurogenesis did not fit the same gradient. In the present study we could identify the neuroepithelial source of the trigeminal and facial motor nuclei in a shared midline matrix (the upper medullary basal plate) and were able to trace the migration of the cells to two separate sites. We were less successful with the retrofacial and ambiguus nuclei. Both appear to derive from the same midline region, but we could not identify with sufficient assurance either their sites of origin or migratory paths.

Cytogenetic zone VE (Altman and Bayer 1980c: Figs. 19, 20) designated the four vestibular nuclei: the lateral, superior, inferior, and medial. We have subdivided this zone into an earlier-generated, large-celled rostral division (VIIIl and VIIIs) and a later-generated, smaller-celled caudal division (VIIIi and VIIIm). In both components there was a lateral-to-medial cytogenetic gradient. In the present embryonic study we could identify the neuroepithelial region which must be the source of the neurons of the vestibular nuclei. In younger embryos (Fig. 38), cells were seen migrating from the alar plate in the direction of the vestibular nerve. This region was presumed to be the source of neurons of the lateral vestibular nucleus and possibly of the superior vestibular nucleus. Subsequently, as a result of the appearance of a new neuroepithelial region over the recess of the fourth ventricle laterally, the vestibular neuroepithelium became displaced medially. By this time two subdivisions could be recognized (Fig. 45): the receding lateral portion with larger, differentiating neurons in its vicinity, and an active medial portion. The latter was identified as the source of neurons of the inferior and medial vestibular nuclei.

9 Summary

In this study we have utilized three sets of anatomic preparations to study three aspects of the morphogenetic development of the medullary cranial nerve system of the rat. In a series of whole-body thymidine radiograms we determined quantitatively the time of origin of neurons of the trigeminal, facial, vestibular, glossopharyngeal, and vagal ganglia. In a collection of embryos we examined the development of these ganglia from the earliest stages until birth. Finally, using previously obtained data about the time of origin of neurons in the sensory and motor nuclei of the medulla directly associated with the cranial nerves, we attempted to correlate the time course of events centrally and peripherally.

The Trigeminal Nerve Complex. The trigeminal ganglion cells are produced between days E11 and E15, with a peak on day E13. The ganglion cells produced on days E14 and E15 were predominantly of the small type. There were no differences in the production times of neurons of the anteromedial and posterolateral lobes, and no cytogenetic gradient was evident in the ganglion as a whole.

The embryonic origins of the trigeminal ganglion were investigated from day E10 onward. The anlage of the trigeminal ganglion was evident by day E11, when two cell aggregates could be distinguished, a lateral (possibly of placodal origin) and a medial (possibly of neural crest derivation). The cytological differentiation of cells of the lateral anlage began on day E13, and by day E15 small and large neurons were scattered in this region. The differentiation of these cells continued, but there was no difference in the stainability of small and large neurons throughout the embryonic and early postnatal periods. In contrast, the cells derived from the medial anlage remained undifferentiated over this period. These cells formed a peripheral boundary cap around the medulla at the site of entry and exit of trigeminal afferents and efferents. The apparent progeny of these cells spread peripherally along the trigeminal nerve but did not penetrate the nerve centrally. It was postulated that the cells of the trigeminal medial anlage are the source of trigeminal Schwann cells.

On day E13, proximal trigeminal afferents began to traverse the trigeminal boundary cap, and trigeminal motor neurons (generated mostly on day E12) began to migrate from a medial position in the medulla laterally. The motor neurons reached the region facing the boundary cap on day E14 and E15, and their efferents began to traverse the boundary cap at the same time. The migration of the neurons of the principal sensory nucleus began from a posterolateral source on day E15, and on the following days these cells became interposed between the trigeminal boundary cap and the motor nucleus. It was postulated that the trigeminal boundary cap serves as a guidepost for the migrating cells and the growing fibers of the lower trigeminal complex. It was also hypothesized that the cephalic flexion of the embryo on day E12, which brings the face area into the neighborhood of the ventral medulla, is a mechanism that facilitates the diffusion of guiding agents from target organs of the face to the growing afferents of the trigeminal ganglion.

The Facial Nerve Complex. The neurons of the facial ganglion are generated between days E11 and E14, with 86% of the cells being produced on days E12 and E13. The anlage of the facial ganglion was identified on day E11. Contrary to earlier claims that the facial ganglion is derived from a shared acousticofacial primordium, a discrete facial anlage was recognizable before the appearance of the anlage of the otic ganglion. The facial ganglion is derived from the ectodermal thickening (placode) in the cleft region between the hyoid arch and the maxillomandibular enlargement. In parallel with their early origin, the cytological maturation of the facial ganglion cells preceded the maturation of the ganglion cells of the other cranial nerve ganglia.

In day E14 embryos, we observed in the medial aspect of the neuroepithelium, at the level of the facial ganglion (rhombomere 3), a collection of flask-shaped cells with laterally oriented fibers. These cells were identified as the premigratory facial motor neurons. Fibers issuing from this region coursed directly toward the facial boundary cap. Some facial efferents by day E14 reached the facial ganglion itself and joined its afferents. On day E15 the facial motor neurons situated medially began to migrate caudally and then ventrolaterally, and settled in their final location on day E15 and E16. By day E17 the facial nerve had acquired its typical looped course.

The Vestibular Nerve Complex. The neurons of the vestibular ganglion are produced (ahead of the spiral ganglion cells) between days E12 and E14, with 97% of the cells being generated on days E12 and E13. In embryos the detachment of cells from the dorsal wall of the otic vesicle was recognized by day E12, and beginning on this day the primordium of the vestibular ganglion became inserted between the otic vesicle caudally and the facial ganglion rostrally. A second stream of cells began to leave the otic vesicle on day E14; these cells were considered to be the precursors of the neurons of the spiral ganglion.

Fibers of the vestibular nerve could be traced into the medulla, by way of the facial boundary cap, beginning on day E13. A stream of cells apparently moving in the same direction from the alar plate was considered, on the basis of thymidine-radio-graphic datings, to be the early-produced neurons of the lateral vestibular nucleus. Two subdivisions of the vestibular neuroepithelium could be distinguished in day E15 rats: a receding lateral zone, presumably generating the earlier-arising large neurons of the lateral and superior vestibular nuclei; and a more active medial zone interpreted as a source of neurons of the inferior and medial vestibular nuclei. A far-lateral, very active neuroepithelial zone over the recess of the fourth ventricle was considered the germinal source of the late-forming neurons of the cochlear nuclei.

The Glossopharyngeal and Vagal Nerve Complex. The neurons of the inferior (or trunk) ganglia of the glossopharyngeal and vagal nerves are generated relatively early and over a protracted period, whereas the neurons of the superior (or root) ganglia of the two nerves are generated later and more rapidly (with about 80% of the cells being produced on day E14). These observations suggest that the inferior ganglia of the two nerves derive from one shared germinal source, and the superior ganglia from another common primordium.

The rudiments of the inferior and superior ganglia were identifiable by day E12. In day E13 rats the root ganglia appeared to be contiguous with epithelial thickenings of the oral cavity, and the trunk ganglia with epithelial thickenings of the pharyngeal cavity. By day E14 the neurons of the inferior ganglia showed signs of cytological differentiation, while the cells of the superior ganglia were still primitive in appearance.

A stream of migrating cells was identified in day E13 and E14 rats between the basal plate rostrally and the glossopharyngeal and vagal boundary caps caudally. On the basis of available datings, it was assumed that these were the early-produced neurons of the retrofacial nucleus. A second stream of cells seen on day E16 may represent the very late-forming neurons of the ambiguus nucleus. In an auxiliary study we examined the development of the hypoglossal nucleus, with reference to the site of origin of its neurons and the course of its efferents. As in the case of the other nerves (mixed or sensory), the fibers of this motor nerve leave the medulla at points where the hypoglossal boundary caps are located.

Comparison of the Different Cranial Nerve Complexes. There are similarities and differences in the time and site of origin of neurons and the routes of cell migration and fiber growth in the different cranial nerve complexes. Our observations suggest that all cranial nerve ganglia derive from placodes, with no contributions made to the neuronal population by the neural crest. However, most of the cranial nerve ganglia originate at different times or with different temporal patterns, which is relatable to the observation that they derive from discrete components of the placodal system. The two notable exceptions are the vestibular and spiral ganglia, which originate sequentially from the otic vesicle, and the glossopharyngeal and vagal ganglia in which the two superior ganglia and the two inferior ganglia have shared origins.

The order of neurogenesis in the cranial nerve ganglia is not matched by the sequence of neuron production in the associated nuclei of the medulla. For instance, the neurons of the facial ganglion are produced before the neurons of the trigeminal ganglion, but the reverse order was obtained for the motor neurons of the facial and trigeminal nuclei. Obviously, the production of ganglion cells peripherally and motor neurons centrally are not only independent events, but are unlikely to be causally linked. However, in most instances the ganglion cells are generated before the neurons of the sensory nuclei of the same complex, which would allow for the exertion of organizing influences by the primary afferents on their central receiving neurons. In all instances the primary afferents of a nerve traverse the nerve-specific boundary cap before penetrating the medulla; the efferents, likewise, leave the medulla by approaching the boundary cap. Apparently, the nerve specific boundary caps (which probably derive from the neural crest) guide the growth of components of each cranial nerve at the interface between center and periphery. Subsequently the boundary cap becomes the source of Schwann cells of the peripheral nerves.

The motor neurons of the cranial nerve nuclei considered in this study all originate from a shared germinal column in the basal plate of the medulla. But each has a different fate afterward. The motor neurons of the trigeminal nerve migrate laterally in the direction of the afferents of the trigeminal ganglion. The motor neurons of the facial nerve send their efferents in the direction of the seventh nerve boundary cap, but the perikarya migrate thereafter caudally and ventrally. The motor neurons of the retrofacial and ambiguus nuclei also migrate caudally, but settle in the midportion of the medulla. Finally, the hypoglossal motor neurons do not migrate at all but settle near the ventricle.

Abbreviations

a	anterior	RE	retrofacial nucleus
AM	ambiguus nucleus	RET	retina
AMm	migrating cells of AM	RH	rhombencephalon
AMz	neuroepithelium of AM	rp	recess plate
ap	alar plate	RP	Rathke's pouch
AQ	aqueduct	SC	superior colliculus
AIII	third aortic arch	SL	nucleus of the solitary tract
AIIIp	placode of AIII	so	somites
AIV	fourth aortic arch	SP	spinal cord
AIVp	placode of AIV	st	solitary tract
BG	basal ganglia	ST	subthalamus
bc	boundary cap	tb	temporal bone
bm	basement membrane	TE	telencephalon
bp	basal plate	TH	thalamus
c	caudal	tg	tongue
CC	cochlear canal	v	ventral
CN	cochlear nuclei	VC	vestibular canal
CNz	neuroepithelium of cochlear nuclei	VEz	neuroepithelium of the vestibular nuclei
CO	cochlea		
CP	cerebellar plate	VEz1	zone 1 of VEz
d	dorsal	VEz2	zone 2 of VEz
DI	diencephalon	vL	lateral ventricle
DZ	differentiating zone (mantle layer)	VN	vestibular nuclei
E	embryonic	VSp	visceral placode
HI	hippocampus	vIII	third ventricle
HT	hypothalamus	vIV	fourth ventricle
HY	hyoid arch	vIVr	recess of vIV
HYp	placode of HY	Vbc	boundary cap of trigeminal ganglion
ic	internal capsule	VG	trigeminal ganglion (or anlage)
IO	inferior olive	VGa	trigeminal ganglion, anteromedial lobe
l	lateral		
m	medial	VGp	trigeminal ganglion, posterolateral lobe
MD	mandibular arch		
ME	medulla	Vi	interpolar nucleus of trigeminal nerve
MM	maxillomandibular enlargement		
MMp	placode of MM	Vla	trigeminal lateral anlage
MS	mesencephalon	Vma	trigeminal medial anlage
MX	maxillary process	Vmd	trigeminal nerve, mandibular arch
NE	neuroepithelium	Vmn	trigeminal nerve, motor
oc	oral cavity	Vmx	trigeminal nerve, maxillary branch
OS	optic stalk	VM	trigeminal motor nucleus
OV	otic vesicle	VMm	migrating cells of VM
p	posterior	VMz	neuroepithelial zone (source) of VM
P	postnatal		
ph	pharyngeal cavity	Vn	trigeminal nerve
pi	pia	Vnc	trigeminal nerve, central part
pl	placode	Vnd	trigeminal nerve, descending
PO	pontine gray	Vnp	trigeminal nerve, peripheral part
PR	prosencephalon	Vo	oral nucleus of trigeminal nerve
PT	pretectum	Vop	trigeminal nerve, ophthalmic branch
r	rostral	Vpl	trigeminal placode

VPR	principal sensory nucleus of tri-geminal nerve	VIIIn	acoustic nerve
VPRm	migrating cells of VPR	VIIIs	superior vestibular nucleus
VPRz	neuroepithelial zone (source) of VM	VIIIvn	vestibular nerve
		VIIIms	vestibular nuclear migratory stream
Vs	spinal nucleus of trigeminal nerve	IXG	glossopharyngeal ganglia
VIIbc	boundary cap of VIIG	IXGi	inferior glossopharyngeal (petrosal) ganglion
VIIG	facial ganglion (or anlage)		
VIImn	facial nerve, motor	IXGs	superior glossopharyngeal ganglion
VIIM	facial motor, nucleus	IXMz	germinal source of retrofacial neurons
VIIMm	migrating (or premigratory) facial motor neurons		
		IX–Xbc	boundary cap of nerves IX–X
VIIMz	germinal source of facial motor neurons	IX–Xmn	motor nerves of IX–X
		XD	dorsal nucleus of vagus
VIIn	facial nerve	XGi	inferior vagal (nodose) ganglion
VIIsn	facial nerve, sensory	XGs	superior vagal (jugular) ganglion
VIIIbc	boundary cap of VIIIG	XMz	germinal source of ambiguous neurons
VIIIG	otic ganglion		
VIIIGs	spiral ganglion	XIn	spinal accessory nerve
VIIIGv	vestibular ganglion	XIIbc	boundary cap of XII nerve
VIIIi	inferior vestibular nucleus	XIIM	hypoglossal motor nucleus
VIIIl	lateral vestibular nucleus (Deiters')	XIIMz	germinal source of hypoglossal neurons
VIIIm	medial vestibular nucleus		
VIIIms	migratory stream of vestibular nuclei	XIIn	hypoglossal nerve

References

Adelmann HB (1925) The development of the neural folds and cranial ganglia of the rat. J Comp Neurol 39:19–123

Allen WF (1924) Localization in the ganglion semilunare of the cat. J Comp Neurol 38:1–26

Altman J, Bayer SA (1978a) Development of the diencephalon in the rat. I. Autoradiographic study of the time of origin and settling patterns of neurons of the hypothalamus. J Comp Neurol 182:945–972

Altman J, Bayer SA (1978b) Development of the diencephalon in the rat. II. Correlation of the embryonic development of the hypothalamus with the time of origin of its neurons. J Comp Neurol 182:973–994

Altman J, Bayer SA (1980a) Development of the brain stem in the rat. I. Thymidine-radiographic study of the time of origin of neurons of the lower medulla. J Comp Neurol 194:1–35

Altman J, Bayer SA (1980b) Development of the brain stem in the rat. II. Thymidine-radiographic study of the time of origin of neurons of the upper medulla, excluding the vestibular and auditory nuclei. J Comp Neurol 194:37–56

Altman J, Bayer SA (1980c) Development of the brain stem in the rat. III. Thymidine-radiographic study of the time of origin of neurons of the vestibular and auditory nuclei of the upper medulla. J Comp Neurol 194:877–904

Altman J, Bayer SA (1980d) Development of the brain stem in the rat. IV. Thymidine-radiographic study of the time of origin of neurons in the pontine region. J Comp Neurol 194:905–929

Altman J, Bayer SA (1981a) Development of the brain stem in the rat. V. Thymidine-radiographic study of the time of origin of neurons in the midbrain tegmentum. J Comp Neurol 198:677–716

Altman J, Bayer SA (1981b) Time of origin of neurons of the rat inferior colliculus and the relations between cytogenesis and tonotopic order in the auditory pathway. Exp Brain Res 42:411–423

Angulo AW (1951) A comparison of the growth and differentiation of the trigeminal ganglia with the cervical spinal ganglia in albino rat embryos. J Comp Neurol 95:53–71

Ariëns Kappers CU (1917) Further contributions on neurobiotaxis. IX. An attempt to compare the phenomena of neurobiotaxis with other phenomena of taxis and tropism. J Comp Neurol 27:261–298

Ariëns Kappers CU, Huber GC, Crosby EC (1936) The comparative anatomy of the nervous system of vertebrates, including man, vol 1. Macmillan, New York (Reprinted, Hafner, New York, 1967)

Arvidsson J (1975) Location of cat trigeminal ganglion cells innervating dental pulp of upper and lower canines studied by retrograde transport of horseradish peroxidase. Brain Res 99:135–139

Bartelmez GW (1922) The origin of the otic and optic primordia in man. J Comp Neurol 34:201–232

Batten EH (1958) The origin of the acoustic ganglion in the sheep. J Embryol Exp Morphol 6:597–615

Beaver DL, Moses HL, Ganote CE (1965) Electron microscopy of the trigeminal ganglion. II. Autopsy study of human ganglia. Arch Pathol 79:557–570

Beckstead RM, Norgren R (1979) An autoradiographic examination of the central distribution of the trigeminal, facial, glossopharyngeal, and vagal nerves in the monkey. J Comp Neurol 184:455–472

Blomquist AJ, Antem A (1965) Localization of the terminals of the tongue afferents in the nucleus of the solitary tract. J Comp Neurol 124:127–130

Bok ST (1915) Die Entwicklung der Hirnnerven und ihrer zentralen Bahnen. Die stimulogene Fibrillation. Folia Neuro-Biol 9:475–565

Boudreau JC, Bradley BE, Bierer PR, Kruger S, Tsuchitani C (1971) Single unit recordings from the geniculate ganglion of the facial nerve of the cat. Exp Brain Res 13:461–488

Brodal A (1981) Neurological anatomy in relation to clinical medicine, 3rd edn. Oxford Univ Press, New York

Bruesch SR (1944) The distribution of myelinated afferent fibers in the branches of the cat's facial nerve. J Comp Neurol 81:169–191

Burton H, Craig AD, Poulos DA, Molt JT (1979) Efferent projections from temperature sensitive recording loci within the marginal zone of the nucleus caudalis of the spinal trigeminal complex in the cat. J Comp Neurol 183:753–778

Buskirk CV (1954) The seventh nerve complex. J Comp Neurol 82:303–326

Campenhout E Van (1935) Origine du ganglion acoustique chez le porc. Arch Biol (Paris) 46:273–286

Car A, Roman C (1970) Déglutitions et contractions oesophagiennes réflexes produites par le stimulation du bulbe rachidien. Exp Brain Res 11:75–92

Carmel PW, Stein BM (1969) Cell changes in sensory ganglia following proximal and distal nerve section in the monkey. J Comp Neurol 135:145–166

Carpenter MB, Hanna GR (1961) Fiber projections from the spinal trigeminal nucleus in the cat. J Comp Neurol 117:117–125

Chamley JH, Dowel JJ (1975) Specificity of nerve fibre "attraction" to autonomic effector organs in tissue culture. Exp Cell Res 90:1–7

Chamley JH, Goller I, Burnstock G (1973) Selective growth of sympathetic nerve fibers to explants of normally densely innervated autonomic effector organs in tissue culture. Dev Biol 31:362–379

Cottle MK (1964) Degeneration studies of primary afferents of IXth and Xth cranial nerves in the cat. J Comp Neurol 122:329–345

Cottle MKW, Calaresu FR (1975) Projections from the nucleus and tractus solitarius in the cat. J Comp Neurol 161:143–158

Coughlin MD (1975) Target organ stimulation of parasympathetic nerve growth in the developing mouse submandibular gland. Dev Biol 43:140–158

D'Amico-Martel A, Noden DM (1980) An autoradiographic analysis of the development of the chick trigeminal ganglion. J Embryol Exp Morphol 55:167–182

Detwiler SR, Kehoe K (1939) Further observations on the origin of the sheath cells of Schwann. J Exp Zool 81:415–435

Dom R, Falls W, Martin GF (1973) The motor nucleus of the facial nerve in the opossum *(Didelphis marsupialis virginiana)*. Its organization and connections. J Comp Neurol 152:373–402

Erzurumlu RS, Killackey HP (1979) Efferent connections of the brainstem trigeminal complex with the facial nucleus of the rat. J Comp Neurol 188:75–86

Foley JO, DuBois F (1973) An experimental study of the facial nerve. J Comp Neurol 79:79–105

Forbes DJ, Welt C (1981) Neurogenesis in the trigeminal ganglion of the albino rat: a quantitative autoradiographic study. J Comp Neurol 199:133–147

Gacek RR (1969) The course and central termination of first order neurons supplying vestibular endorgans in the cat. Acta Oto-Laryng [Suppl] 254:1–66

Gaik GC, Farbman AI (1973a) The chicken trigeminal ganglion. I. An anatomical analysis of the neuron types in the adult. J Morphol 141:43–56

Gaik GC, Farbman AI (1973b) The chicken trigeminal ganglion. II. Fine structure of the neurons during development. J Morphol 141:57–75

Gregg JM, Dixon AD (1973) Somatotopic organization of the trigeminal ganglion in the rat. Arch Oral Biol 18:487–498

Gunn CG, Sevelius G, Puiggari MJ, Myers FK (1968) Vagal cardiomotor mechanisms in the hindbrain of the dog and cat. Am J Physiol 214:258–262

Halley G (1955) The placodal relations of the neural crest in the domestic cat. J Anat 89:133–152

Hamburger V (1961) Experimental analysis of the dual origin of the trigeminal ganglion in the chick embryo. J Ex Zool 148:91–124

Hamilton WJ, Boyd JD, Mossman HW (1964) Human embryology, 3rd ed. Williams and Wilkins, Baltimore

Harrison RG (1924) Neuroblast versus sheath cell in the development of peripheral nerves. J Comp Neurol 37:124–205

Hörstadius S (1950) The neural crest. Oxford Univ Press, London

Iwata N, Kitai ST, Olson S (1972) Afferent component of the facial nerve: its relation to the spinal trigeminal and facial nucleus. Brain Res 43:662–667

Jacobson M (1978) Developmental neurobiology, 2nd edn. Plenum Press, New York

Johnston MC (1966) A radioautographic study of the migration and fate of cranial neural crest cells in the chick embryo. Anat Rec 156:143–155

Kerr FWL (1962) Facial, vagal and glossopharyngeal nerves in the cat. Arch Neurol 6:264–281

Kerr FWL, Lysak WR (1964) Somatotopic organization of trigeminal-ganglion neurons. Arch Neurol 11:593–602

Kimmell DL (1941) Development of the afferent components of the facial, glossopharyngeal and vagus nerves in the rabbit embryo. J Comp Neurol 74:447–471

Kugelberg E (1952) Facial reflexes. Brain 75:385–396

Lawn AM (1966) The localization, in the nucleus ambiguus of the rabbit, of the cells of origin of motor nerve fibers in the glossopharyngeal nerve and various branches of the vagus nerve by means of retrograde degeneration. J Comp Neurol 127:293–306

Lawson SN, Caddy KWT, Biscoe TJ (1974) Development of rat dorsal root ganglion neurons. Studies of cell birthdays and changes in mean cell diameter. Cell Tissue Res 153:399–413

Lindquist C (1973) Reflex organization and contraction properties of facial muscles. Acta Physiol Scand, [Suppl] 393:1–15

Loewy AD, Burton H (1978) Nuclei of the solitary tract: Efferent projections to the lower brain stem and spinal cord of the cat. J Comp Neurol 181:421–450

Mazza JP, Dixon AD (1972) A histological study of chromatolytic cell groups in the trigeminal ganglion of the rat. Arch Oral Biol 17:377–387

Morest DK (1967) Experimental study of the projections of the nucleus of the tractus solitarius and the area postrema in the cat. J Comp Neurol 130:277–300

Moses HL (1967) Comparative fine structure of trigeminal ganglia, including human autopsy studies. J Neurosurg [Suppl] 26:112–126

Narayanan CH, Narayanan Y (1978) Determination of the embryonic origin of the mesencephalic nucleus of the trigeminal nerve in birds. J Embryol Exp Morphol 43:85–105

Noden DM (1975) An analysis of the migratory behavior of avian cephalic neural crest cells. Dev Biol 42:106–130

Nomura S, Mizuno N (1981) Central distribution of afferent and efferent components of the chorda tympani in the cat as revealed by the horseradish peroxidase method. Brain Res 214:229–237

Patten BM, Carlson BM (1974) Foundations of embryology, 3rd edn. McGraw-Hill, New York

Peach R (1972) Fine structural features of light and dark cells in the trigeminal ganglion of the rat. J Neurocytol 1:151–160

Piatt J (1945) Origin of the mesencephalic V root cells in *Amblystoma*. J Comp Neurol 82:35–53

Pineda A, Maxwell DS, Kruger L (1967) The fine structure of neurons and satellite cells in the trigeminal ganglion of cat and monkey. Am J Anat 121:461–488

Rhoton AL (1968) Afferent connections of the facial nerve. J Comp Neurol 133:89–100

Roman C, Car A (1967) Contractions oesophagiennes produites par la stimulation du vague ou du bulbe rachidien. J Physiol (Paris) 59:377–398

Sims TJ, Vaughn JE (1979) The generation of neurons involved in an early reflex pathway of embryonic mouse spinal cord. J Comp Neurol 183:707–720

Spemann H (1938) Embryonic development and induction. Yale Univ Press, New Haven

Stein BM, Carpenter MB (1967) Central projections of the vestibular ganglia innervating specific parts of the labyrinth in the rhesus monkey. Am J Anat 281–318

Stewart WA, King RB (1963) Fiber projections from the nucleus caudalis of the spinal trigeminal nucleus. J Comp Neurol 121:271–286

Streeter GL (1906) On the development of the membranous labyrinth and the acoustic and facial nerves in the human embryo. Am J Anat 6:139–165

Szabo T, Dussardier M (1974) Les noyaux d'origine du nerf vague chez le mouton. Z Zellforsch 63:247–276

Tanaka T (1977) Synaptic activation of facial afferents on the facial neurons of the cat. Brain Res 123:378–383

Tanaka T, Yu H, Kitai ST (1971) Trigeminal and spinal input to the facial nucleus. Brain Res 33:504–508

Tello JF (1922) Les différenciations neuronales dans l'embryon des poulet, pendant les premiers jours de l'incubation. Trab Lab Invest Biol, Univ Madrid 21:1–93

Theisen CT (1979) Effects of hydroxyurea during final neuronal DNA synthesis in dorsal root ganglia of rats. Dev Biol 69:612:626

Tokunaga A, Oka M, Murao T, Yokio H, Okumara T, Hirata T, Miyashita Y, Yoshitatsu S (1958) An experimental study of facial reflex by evoked electromyography. Med J Osaka Univ 9:397–411

Torvik A (1956) Afferent connections to the sensory trigeminal nuclei, the nucleus of the solitary tract and adjacent structures. An experimental study in the rat. J Comp Neurol 106:51–142

Truex RC, Carpenter MB (1969) Human neuroanatomy, 6th edn. Williams and Wilkins, Baltimore

Weigner K (1901) Bemerkungen zur Entwicklung des Ganglion acustico-faciale und des Ganglion semilunare. Anat Anz 19:145–155

Weston JA (1970) The migration and differentiation of neural crest cells. Adv Morphogen 8:41–114

Windle WF (1933) Neurofibrillar development in the central nervous system of cat embryos between 8 and 12 mm long. J Comp Neurol 55:643–723

Yntema CL (1937) An experimental study of the origin of the cells which constitute the VIIth and VIIIth cranial ganglia and nerves in the embryo of *Amblystoma punctatum*. J Exp Zool 75:75–101

Yntema CL (1943) Deficient innervation of the extremities following removal of the neural crest in *Amblystoma*. J Exp Zool 94:319–343

Subject Index

Other Reviews of Interest in This Series

Volume 64
A. Brodal, K. Kawamura
Olivocerebellar Projection: A Review
1980. 45 figures. VII, 140 pages
ISBN 3-540-10305-8

Volume 65
E. Pannese
The Satellite Cells of the Sensory Ganglia
1981. 30 figures. IX, 111 pages
ISBN 3-540-10219-1

Volume 66
H.-M. Schmidt
Die Artikulationsflächen der menschlichen Sprunggelenke
1981. 45 figures, 2 tables.
VII, 81 pages
ISBN 3-540-10306-6

Volume 67
H. Wolburg
Axonal Transport, Degeneration and Regeneration in the Visual System of the Goldfish
1981. 28 figures. IX, 94 pages
ISBN 3-540-10336-8

Volume 70
W. Pfaller
Structure Function Correlation on Rat Kidney
Quantitative Correlation of Structure and Function in the Normal and Injured Rat Kidney
1982. 23 figures. approx. 41 tables.
Approx. 120 pages
ISBN 3-540-11074-7

Volume 71
L. Thuneberg
Interstitial Cells of Cajal: Intestinal Pacemaker Cells?
1982. 94 figures, 1 table. Approx. 120 pages
ISBN 3-540-11261-8

Volume 72
H. Breuker
Seasonal Spermatogenesis in the Mute Swan (Cygnus olor)
1982. 30 figures, 1 table.
Approx. 120 pages
ISBN 3-540-11326-6

Volume 73
G. A. Zweers
The Feeding System of the Pigeon (Columba livia L.)
1982. Approx. 45 figures, approx. 3 tables.
Approx. 90 pages
ISBN 3-540-11332-0

Springer-Verlag
Berlin
Heidelberg
New York

A. G. Brown

Organization in the Spinal Cord

The Anatomy and Physiology of Identified Neurones

1981. 148 figures. XII, 238 pages
ISBN 3-540-10549-2

Contents: Spinal Cord Organization: An Inroduction. – Axons Innervating Hair Follicle Receptors. – Axons Innervating Rapidly Adapting Mechanoreceptors in Glabrous Skin. – Axons Innervating Slowy Adapting Type I Mechanoreceptors. – Axons Innervating Slowly Adapting Type II Mechanoreceptors – Spinocervical Tract Neurones. – Relationships between Hair Follicle Afferent Fibres and Spinocervical Tract Neurones. – Neurones with Axons Ascending the Dorsal Columns. – Other Dorsal Horn Neurones. – The Organization of the Dorsal Horn. – Afferent Fibres from Primary Endings in Muscle Spindles. – Afferent Fibres from Golgi Tendon Organs. – Afferent Fibres from Secondary Endings in Muscle Spindles. – Relationships between Group IA Afferent Fibres and Motoneurones. – Appendix 1: Methods. – Appendix 2: Nomenclature. – References. – Subject Index.

The successful introduction of Horseradish Peroxidase (HRP) as a neurone marker has generated a mass of new data on the spinal cord, leading to a considerable revision of previously accepted neurophysiological ideas. Although some results from the use of this method have been published in scientific journals, the restrictions they impose on space, format and numbers of illustrations led the author of this monograph to undertake its publication to allow extended coverage and discussion of results in one place.

Organization in the Spinal Cord is not just a text on spinal cord anatomy and physiology, it is also a description of results obtained using the intracellular injection of HRP, together with other complementary experiments on selected axons and neurones, specifically on the larger cutaneous axons and dorsal horn neurones and on the larger muscle afferent fibers and motoneurones. This book presents a sufficient amount of new material for review and assessment to suggest that the time has come for a complete rethinking of structural-functional relationships in the mammalian central nervous system. *Organization in the Spinal Cord* will help make neurobiologists aware of the value of working with identified neurones, while providing anatomists, physiologists, pharmacologists and developmental biologists with invaluable information on normal structural-functional relationship in the spinal cord.

Springer-Verlag
Berlin
Heidelberg
New York